approaches to ENVIRONMENTAL STUDIES

General Editor: Gordon A. Perry book 22

Power supplies

Leon Metcalfe

BLANDFORD PRESS

POOLE DORSET

Acknowledgments

The publishers are indebted to the following copyright holders for permission to use the illustrations indicated:

Aerofilms Ltd, Fig. 6:18
British Gas Corporation, Figs. 5:12, 5:13, 6:8, 6:9, 6:12, 6:13, 6:22, 6:23
British Tourist Authority, Fig. 4:1
British Waterways Board, Fig. 4:13
Camera Press Ltd, Figs. 1:9, 1:11, 2:2, 3:1, 3:16, 8:4
CEGB, Figs, 4:9, 4:10, 7:2, 7:3, 7:12, 7:13, 7:19
Charrington, Gardner, Locket and Co. Ltd, Figs, 2:4, 5:9
CNES, Figs. 8:4, 8:5
Electricity Council, Figs. 1:3, 1:4, 7:14, 7:15, 7:16, 7:17, 7:18
French Government Tourist Office, Fig. 4:7
Gilbert Gilkes Gordon and Co. Ltd, Fig. 4:11
C. F. Hickling, Fig. 4:6
Keystone Press Agency, Fig. 7:1
Mansell Collection, Figs. 6:2, 7:11
L. E. Metcalfe, Figs. 1:1, 1:2, 1:6, 1:7, 1:8, 1:10, 2:5, 3:5, 3:6, 3:7, 3:9, 3:10, 3:11, 3:12, 4:2, 4:3, 4:8, 5:6, 5:7, 5:8, 5:10, 5:11, 6:14, 6:15, 6:17, 6:19, 6:20, 6:21, 7:5, 7:6, 7:7, 7:8, 7:9
Marion Mills, Figs. 3:17, 4:4, 8:2, 8:3
National Coal Board, Figs. 1:5, 5:3, 5:4
NFBC, Fig. 4:12
National Maritime Museum, Fig. 3:3
NOAA, Figs. 3:13, 3:14
Orkney Public Library, Kirkwall, Fig. 3:8
Portman Artists, Figs. 2:1, 2:3, 3:4, 5:5
Shell Photo Library, Figs. 6:3, 6:4, 6:5, 6:6, 6:7
Solarfilms, Fig. 4:5
UKAEA, Fig. 7:4
Yachts and Yachting, Fig. 3:15

First published in 1975

Link House, West Street, Poole, Dorset, BH15 1LL

ISBN 0 7137 3496 5

Printed and bound in Great Britain by Jarrold & Sons Ltd, Norwich

Contents

Editor's Note

This book is part of a wider scheme, *Approaches to Environmental Studies,* published by Blandford Press. This consists of class books, programmed texts, teachers' books and ancillary material. Programmes, involving the study of the environment and not requiring teaching machines, have been written by Colin Kefford. These also are available from Blandford Press.

Specially prepared illustrations in the form of 2 × 2 inch slides in colour to help in these studies may be purchased from: The Slide Centre Ltd, Portman House, 17 Brodrick Road, London, S.W.17.

Teachers are recommended to use (*a*) the Handbook, and (*b*) Guide No. 4 which supports and expands the ideas in the class books and introduces other areas of experimentation and investigation.

G.A.P.

1:1 Bicycles allow us to use the power through our muscles more efficiently than walking.

1:2 Very little human energy is used to drive a car. Petrol is being burnt to produce the power to propel the car. Notice the power station in the background.

1-OUR NEED FOR POWER

For transport

How did you travel to school this morning? Perhaps you walked or cycled. You may have travelled by bus, train or car. It is certain that by whichever means you did arrive at school, a form of power was used. If you walked, you made use of your own muscle power, as men have done for thousands of years. Your cycle is also propelled along the road by the power of your muscles, but you get to school quicker with this method because the cycle is a machine which has been developed by man to make his muscle power more efficient.

The train, bus and car are even more effective than the cycle because the amount of muscle power involved in getting you to school by these is reduced to nothing. The driver of each form of transport has to use his muscle power to control the vehicle, but you, as a passenger, are able to relax while you are carried along to your destination.

For light

Was it dark when you got up this morning? If it was not, it probably will be before you go to bed tonight. When the house is dark you will expect to flick a switch for an electric light to come on to enable you to continue seeing after the world's natural light, the sun, has disappeared over the horizon.

For heat

Are you warm at the moment? I am sure you are, because if the sun is not warming you, the heating system in the building will be doing the job efficiently. You may have central heating at home or at school, or you may rely on 'open' fires. Whichever has been installed you are almost certainly using coal, gas, oil or electricity to maintain your body at a comfortable temperature.

For cooking

Have you had anything to eat today? I hope so, because your muscles need the food you eat to help to keep them working properly. How was your food cooked and your drinks heated? Once again, you need gas, oil, electricity or perhaps even coal to help to do these jobs.

1:3, 1:4, 1:5 Much of the power used around the home is taken for granted. We expect it to be there and give no thought to how it comes. But the old-fashioned kitchen range (*above left*) needed a lot of effort to keep it supplied with coal and had to be carefully tended.

For recreation

Are you going to listen to a record, tape or radio, or watch television today? I expect you hope to, but what happens if there is no electricity available?

Power in our lives

Have you ever thought what our lives would be like without power? It would be impossible to live anything like the lives we lead now without artificial light, heat and transport. But these are only a few examples of the ways in which our present-day lives depend on many different sources of power. We have not mentioned rockets to the planets, aeroplanes flying round the world, machines in factories making the things we need, or even the mechanical aids that are used to build the factories to make the things, and the powered implements the farmer uses to plant, grow and harvest much of the food we need to eat to stay alive.

The search for new power sources

It certainly seems that life would be impossible without our wide range of available power, but there was once a time when man had only his muscles and no other power to help him. As thousands of years have gone by, man, in order to improve his way of life, has been forced to search for better ways of doing things. No progress or advance in civilisation would have been possible without these efforts. The search for more and better forms of power still goes on today for the same reasons as of old plus one other important reason. This last reason is an extremely serious one: nearly all the means of making power which we now have are very quickly being depleted, and there will come a time when they will no longer be available to us. When that time comes, and if we have not found new sources of power, progress for man will be at an end.

The earth's crust

What are the present sources of power, and how can their supplies be exhausted? Some, such as gas, coal and oil, have been mentioned already. Each of these is taken out of the earth's crust and burnt in one form or another to warm us, to give us light, to transport us to different places or to provide the power to make things for us. The unfortunate thing is that they can only be used once. When that shovel of coal is placed on the fire

1:6, 1:7, 1:8 The wind can be felt but not seen unless it is acting upon objects such as the three illustrated here.

and burnt away, its usefulness is over. When that gallon of petrol fuel oil has been pumped into the tank of our car and the car has burnt it while carrying us on our journey, it is no longer available for use as power again. At tea-time it is easy for us to use the last spoonful of sugar out of the sugar bowl, get a fresh packet and refill the bowl. But if you imagine the earth as a huge container holding supplies of gas, coal and oil which we are quickly using up, then it is not difficult for you to realise that there is no fresh packet to open when the earth's 'fuel' container

is empty. 'When it's gone, it's gone', and all the equipment and machinery which need these things to make them work will be useless. This, then, is the important reason why our scientists are busily searching for fresh sources of power.

Electricity

Electrical power is a very important source of power for us. Electricity cannot be mined or pumped out of the earth's crust. It has to be made; and in order to make it we often need to burn coal, gas or oil. Thus electricity made in these ways will no longer be available to us when the three natural sources run out.

Wind and water

Fortunately, we know other ways to make electricity which have not been mentioned yet. Two methods which can be used are basic sources of power which man has been using for perhaps thousands of years, though not for making electricity. These sources are the wind and water power. There is no likelihood of these being exhausted, but that does not mean that there are no problems attached to these forms of power production. The wind is very unreliable: sometimes there is virtually no air movement at all, and at other times the wind blows with such force that it can cause great harm and do much damage.

Water as a power source can be made more useful and reliable by the building of huge dams to hold back a great store of water, so that it can be allowed to flow at a speed and time to suit man and thereby make his electricity. The problem, however, is that sufficient running water, adequate rainfall and suitable land formations do not always occur near the places where a high demand for electricity exists.

1:9, 1:10 Murchison Falls on the River Nile (*left*) and Swallow Falls, Wales (*right*). There is beauty in a waterfall. When watching one, consider how much uncontrolled energy is going to waste.

1:11 Mountaineering and other sports use the power of our muscles. This power is provided through the food we eat.

Nuclear power

Electricity can also be made in nuclear power stations, and this is one reliable source of power which should be developed and improved, so that eventually it could replace to a large extent the other sources which we are using up. However, nuclear power can never entirely solve our future power supply problem, so if we are to continue to develop our civilisation, our scientists must find other means of creating power before our supplies of natural coal, gas and oil are completely exhausted.

Discovery Assignment 1

1 Take a census among the children at your school to find out how they travel to school. Represent this information in the form of a graph.

2 Find a safe cycling area, the playground or playing field, and mark off a hundred metres length. You will need a stopwatch to time the following means of travelling this distance: (*a*) walking, (*b*) running, (*c*) cycling. Perhaps one of your teachers will drive his/her car this distance so that this can be timed also. What can you discover from the results?

3 Make a list of everything you do or use during the day, from getting up to going to bed, which relies upon a power supply other than your own muscles. Some are discussed in this chapter.

4 Imagine that you lived in very early times when you had to rely on your own strength to stay alive. Write a story about your life.

5 Many inventors have tried to make things work by 'perpetual motion'. Find out what this phrase means and why everyone has found it an impossible task.

2-MAN AND ANIMAL POWER

EARLY MAN

Many thousands of years ago when man first walked the earth, his every job had to be done by his own strength alone. His main occupation was to stay alive, to capture and kill enough meat for himself and his family and to gather fruit and berries, while all the time remaining aware of the dangers around him. These dangers came from other groups of men or from any of the meat-eating creatures which wandered the earth. Survival was then man's major concern in life. He had no settled home and few possessions. He knew that he had to carry with him everything he owned as he wandered about from one hunting-and-gathering area to another.

Man, the hunter

The strength he possessed and his ability to plan and think were the major things in his favour during this period of survival. He was not as strong as many of the creatures he killed for food, but his brain gave him the ability to outwit all other animals, and he therefore became a skilled hunter.

Man, the cultivator

As years passed man realised that a lot of his life and energy was being wasted in wandering. He realised that he could settle in one place where there was somewhere to shelter, in a cave perhaps, with running water to drink from a stream or river nearby, and where the land around could be cultivated for the planting of seeds, which would give him easily acquired food at harvest time.

Muscle power

In these periods of early man, all work had to be done by human muscle power, which meant that when food was plentiful the men grew strong and healthy and could work hard. But even in these ideal conditions muscles grow tired until the effort needed for the job becomes too great, and the body must be rested.

The discovery of fire

Another important source of power for early man was fire.

Man must surely have seen natural fire in the form of red-hot lava from volcanoes or lightning striking a tree, causing it to burst into flames. He would, of course, be petrified by the red and yellow licking flames burning and destroying almost everything in their path, but little by little man realised that a controlled fire could be useful to him in keeping him warm and keeping the wild animals, who never overcame their fear of fire, at bay. Eventually he discovered that he could make fire whenever he wanted to by heat created by the friction of two pieces of wood being rapidly rubbed against each other, or by making a spark by striking two flints to set light to dry grass or leaves.

Fire and its uses

With fire at his disposal, early man found other uses for it apart from safety and warmth. He could now cook food and bake wet clay to make bowls and dishes. His arrow and spear tips could be made extremely sharp and strong by charring the outer surface of them in the embers of his fire. When he was making

2:1 Can you work out the forms of power used by these cave-men?

2:2 Life became a little easier for man when he trained animals to use their muscles on his behalf. Can you think of other animals which work for man?

things out of wood – a plough or canoe – he could save himself a lot of effort by allowing the consuming flames of his controlled fire to burn away the parts of the wood he did not require.

Animal power

It was again a big step forward, in the world of power, when man first realised that many of the wild creatures around him could be tamed, and their much greater muscle power applied to many of the jobs he had to do. In many different parts of the world various animals were trained to help with man's work. These included horses, oxen, camels, yaks, elephants, donkeys and dogs. With these creatures coming to his aid, man no longer had to pull his own plough, draw his own water, clear trees from his land or pull his possessions from place to place.

In many countries throughout the world these animal power suppliers still work from dawn till dusk, from birth to death, to make man's life easier.

Mass muscle power

As men wandered less they began to form communities where quite large groups would settle together. When this happened they not only lived together but also worked together, which meant that if there was a tree which one family group and its animals could not move, the community would provide mass muscle power and thereby accomplish much bigger and more difficult tasks.

Muscle power for building

When man became a builder, at certain times it was necessary for thousands of men to use their muscles as a team to create such amazing places as the great stone circles of Stonehenge and Avebury in this country and the pyramids in Egypt. The pyramids, amazing as they are, were built by a power which does not display the best in human beings. The ancient Egyptians lived in a very advanced civilisation, and they became very powerful conquerors of the people around them. When a tribe or nation was conquered, its people were expected to become the slaves of the conquerors. Consequently the beautiful buildings and pyramids of ancient Egypt were built largely by the muscle power of thousands of slaves.

Brain power

They were aided, to some extent, by the intelligent engineers of the time, who realised that their huge blocks of building stone could be moved more easily by fewer men and muscles if certain plans and preparations were first made. For example, the Egyptians made use of the wheel on their chariots and wagons and, although they could not build a wagon to support the heavy weight of their building blocks, they could use tree trunks as rollers on which the blocks could be placed and moved more quickly by fewer men. To get the blocks raised sufficiently to put the rollers underneath they would put a lever under one end of the block, and again they would find that fewer slaves would be required to raise the block than if it had to be lifted bodily.

Another method employed to make the slaves' muscular efforts more efficient was when a block needed to be stood on end: a long sloping ramp was built, up which the block could be moved on its rollers rather than being lifted vertically into

2:3 Although the builders of Stonehenge used levers to erect the stones, much of the power required came from the men working together.

position by men alone. The distance travelled by the block was greater, but the effort required was far less.

It is amazing to realise that the ancient Egyptians worshipped many different animals as gods and yet put their fellow men into such hard-working slavery. Slavery was not restricted to the ancient civilisations. Men right up to modern times have misused their fellow human-beings in very similar ways, keeping them alive in terrible conditions, for the sole reason that they needed to make use of the power of their muscles.

Even today many parts of the world are less well-developed in their use of power and machines than we are. In these countries, for example China and India, the power from the muscles of men and animals is still one of the most important means by which are accomplished such great feats of civil engineering as the digging of major canals and the building of huge dams. In some respects these nations are fortunate in that they have huge populations, and therefore the manpower is available to take on these immense tasks.

2:4, 2:5 Animal power for pleasure and work. Compare this coal wagon with the one on page 50.

Discovery Assignment 2

1 Collect information and pictures to make a frieze for your classroom to show (*a*) man making use of his own muscles, alone or with many others, and (*b*) man using animal power.

2 This frieze can be enlarged to include more pictures and information about the many huge monuments which man has built with muscle power. There are some ideas for this in the chapter but there are many more to discover; for example, the Incas of Peru and the building of their famous cities. Think also of the Greeks and Romans.

3 Try pushing heavy weights yourself and with the help of others. Why were the inventions of the wheel, the lever and the pulley important?

4 Can you imagine what it was like being an Egyptian slave building the pyramids, or a Roman slave rowing in a galley? More recently you could have been a slave in the southern states of America working on a cotton plantation. Choose one of these slave lives and write about your existence in such conditions.

3-THE WIND

We are all familiar with the winds on our planet, and we take them very much for granted. We know that they are there because we can feel them, sometimes light and pleasantly cooling on a hot summer's day, sometimes frightening in their power, blowing the clouds across the sky and shaking the trees out of the ground by their roots. We accept them because we cannot control them, and man has only been able to make use of them in limited ways.

The 'windless' planet

It would, however, be a very different world if we had no winds, a situation which confronted the American astronauts who first landed on the moon in their spaceship, Apollo 11. They wanted to erect and fly the American flag – the stars and stripes – but they knew that in the airless conditions of the moon it would not fly. They therefore had to hold it open by means of a spring!

3:1 The 'windless' planet. The Stars and Stripes had to be kept aloft by means of a spring.

The movement of air

The reason for our world having winds blowing on its surface and the moon having none is that the earth has an invisible cloak of air, which the moon does not have. The creation of wind, or the mass movement of air, is a very complicated study, but in its simplest form it can be explained by taking the air in a room as an example. When the weather is cold, human beings make efforts to warm the building in which they live or work. This usually means they light a gas, coal, oil or electric fire or have radiators in the room. All these sources of heat affect the air around them by warming it up. The molecules making up the air move faster as they are heated, and they spread out. As they disperse they become lighter and, therefore, rise towards the ceiling of the room. Their place, close to the source of heat, must be filled, so cooler air flows into the space being vacated by the rising air. Thus we have a movement of air, or a wind as we would call it outdoors.

Wind systems of the world

The major source of heat for the world comes from the sun and, as you might expect, some areas of the earth's surface get more direct sunlight and heat from the sun than others. The extremes are, of course, the areas around the equator, which receive a lot of heat from the sun, compared with the polar ice caps in the Arctic and Antarctic. Because different parts of the surface of the earth heat up at different rates (land heats quicker than sea masses), things become more complicated. Another confusing aspect of our wind system is the fact that the earth is spinning round on its axis, so that different parts of the earth are being heated at different times, and the moving air is affected by the rotation of the earth. You can find out much more about the major wind systems of our planet in Book 4 of the Environmental Studies series, *The Air We Need.*

Using the wind for powering ships

Man's first attempts to harness the power of the wind came when he realised that his boats and canoes could be helped across the surface of the water when the wind was blowing in the direction he wanted to travel. By erecting a mast and a square of cloth, he could use the wind to propel his boat along. The difficulty

3:2, 3:3 The Chinese junk still works along the rivers and coasts of China but the warship has long since disappeared from the seas.

with this, however, was that with a square sail fixed to a mast he could only use it when travelling *with* the wind. It was not until the 15th century that he discovered that it was possible with the use of a 'jib' sail to travel, not directly against the wind, but to tack or zig-zag into it. The jib sail is a triangular sail which is attached to the foremast and the bowsprit (the pole extending directly out in front of the ship). This type of sail catches the wind across the ship and balloons out to propel the vessel forward. It must be remembered, however, that no matter how much sail a ship carried, it could not travel across the seas or oceans when there was no wind. Even the magnificent sailing ships of the 19th century would often be 'becalmed' for days on end waiting for the wind to rise.

Although in some parts of the world, such as the Indian Ocean and the China seas, sailing ships are still used commercially for cargo carrying, the majority of sea-going vessels are today powered by diesel engines. Sailing boats are still very popular throughout the world's seas, oceans, lakes and rivers for pleasure and recreation.

WINDMILLS

No one really knows when 'sails' were first used on land to make the power of the wind work for man. A form of windmill was certainly in use in Persia in the 10th century though some reports state that they were first used in China. Records in this country show, however, that the first very small and inefficient windmill was in use about the year 1190. From this date through to the middle of the nineteenth century, the windmill was developed and improved, new types were introduced and their efficiency increased until there were about 10,000 working windmills in this country alone. Most people agree that it was in Britain that the finest windmills were designed and built. The end of the useful life of the windmill was brought about when man discovered new and more reliable sources of power such as coal, which was burnt to heat water and produce the steam that powered the machinery of the new mills.

3:4 Cutaway view of a post mill at work, showing the moving parts: 1 windshaft, 2 brake wheel, 3 wallower, 4 millstones (runner stone and bed stone shown without casings), 5 sack hoist, 6 post, 7 quarter bars, 8 cross trees, 9 vanes.

3:5

3:6

3:5, 3:6, 3:7 Many windmills have completely disappeared but some, like the tower mill in 3:5, remain standing but derelict. Others have been rescued and restored by local councils or societies. The post mill in 3:6 has been partly restored and the one in 3:7 is now in working order. Notice the tail wheel on the left of the picture.

3:7

The post mill

It is generally agreed that the earliest successful windmill in this country was of the post mill type. In this form almost the entire mill was supported on a strong oak post at its centre. The whole mill was designed to turn on this centre post so that the sails could be turned to catch most of the wind. The miller entered his post mill by means of a ladder attached to the body of the mill opposite the sails. Through the steps of this ladder came the tail-pole, which was a long lever used by the miller to turn his mill and sails into the wind. The ladder was very strongly made since it acted as a steady for the whole mill when the wind was gusting fiercely.

The smock mill

During the 17th century a rather more complicated mill was designed and introduced; it was called a smock mill. People say it got its name because it looked like a man dressed in a peasant's smock. The big difference between this type of mill and the earlier post mill was that only the cap or topmost part of the mill carrying the sails needed to be turned to use the power of the wind. The whole of the tower and machinery beneath the cap remained in one fixed position.

The tower mill

Smock mills, like post mills, were mainly constructed of wood, but later another type of mill was introduced which was basically similar to the smock mill though the tower was constructed of brick or stone. They were called tower mills and they proved to be very strong and long-lasting.

3:8 Notice the difference between this typical Orkney windmill and the English mills on page 21.

The fan-tail

With the introduction of smock and tower mills an improved method of turning the sails to the wind was designed. This was a fan-tail, a rudder-like mechanism on the opposite side of the cap to the sails, and this automatically adjusted the direction of the sails as the wind changed. It worked in a similar way to a weather vane pointing its arrow constantly into the wind. A variation of this fan-tail was produced to help the owners of post mills, but in these cases the fan-tail mechanism was supported and run on wheels on a track laid round the outside walls at the foot of the mill.

The sails

Whatever type of windmill was in use through the centuries, one thing which they all had in common was a set of sails. These, as we shall discover, were improved and developed as time passed, but the earliest known sails were simply frameworks of wood covered with cloth. The miller could furl or roll these cloth sails up if he wished so that there was no danger of a sudden gust of wind turning or damaging the sails when they were not in use. Similarly, if there was a very strong wind blowing and the miller did not want to use all its power, then he could partly furl the cloth to reduce the wind's effect on his sails.

The spring sail

In 1772, a Scottish millwright called Andrew Meikle invented a new type of sail called a spring sail. These were made up of a series of spring-loaded shutters, rather similar to a Venetian blind, which could be adjusted through various positions from fully open – to use the least amount of wind – to fully closed where the wind could strike on the full area of the sail. The miller could set the tension on each sail so that the shutters would adjust themselves by opening slightly if the mill was caught in a sudden gust of wind. The shutters were usually made of wood, but these proved to be very heavy, so attempts were made to make them of a wooden framework covered with canvas.

The self-reefing sail

In 1807 Sir William Cubitt invented a patent self-reefing sail which enabled the miller, for the first time, to adjust his sails

3:9 (*left*) An agricultural wind pump raising water to be used by the stock in the fields.

3:10 (*top right*) This is one of ten wind pump models kept as a hobby by an enthusiast.

3:11 (*below right*) This ornate wind pump raises water from a well to supply a private lake.

from inside the mill without stopping the mill whilst he did it. It was quite a simple mechanism which attached all the shutters on each sail to a 'spider' – an armed device fixed at the junction of the sails. The controlling arm, or the 'striking rod' as it was called, ran from the spider into the mill.

How many sails?

The number of sails used on windmills was not always four. Some had five, six or eight sails. The only remaining eight-sail mill in England is at Heckington in Lincolnshire. It is a tower mill built in 1830. The most inconvenient number of sails was five because in the case of the other mills if one sail was damaged, its opposite sail could be brought out of use and the mill could continue work and still be in balance. With the five-sailer this was never possible. Mills were built with more and more sails because it was found that the operation of the mill was smoother as less vibration resulted when the number of sails was increased.

The work of a windmill – grinding grain

By far the majority of all wind-powered mills were used for grinding corn into flour, so somewhere within the body of the mill were housed the millstones. Let us follow a sack of grain through a typical mill to find out more about the working parts and the heart of the windmill. First of all, it was necessary to hoist the grain in its sack to the top of the mill where the miller emptied it into a grain bin. From here the grain slid down a chute to a hopper positioned above the millstones. With the

3:12 This sailing dinghy is using wind power for pleasure.

3:13 Hurricanes leave destruction in their wake.

3:14 This picture shows the damage that uncontrolled fury of a hurricane can create.

gears engaged between the sails and the millstones, the mill began operations. The grain dropped steadily from the hopper on to a feed shoe, as it was called, which was vibrated to encourage the grain to feed into the eye or centre of the upper stone. Only the upper runner stone revolved, the second stone, the bed-stone, remaining stationary. As the stone turned, the grain was ground to flour as it passed out from the centre to the edge of the stones. Here the flour appeared and fell to the mill floor below.

3:15 Land yachting is a growing wind-powered sport.

Draining land

In Holland and the area of England called the Fens, many windmills were used for draining the land of surplus water. The wind turned the sails, which in turn rotated a shaft attached to a paddle wheel. As the paddle wheel was turned round and round, it scooped up water from a low level and emptied it out much higher up into drainage channels from which it could flow away.

Wind power today

Nowadays in Britain the windmill is very little used. Some still remain, restored for us to visit, but are not used commercially. In Holland they still pump the land dry of surplus water, but in the main the only use present-day people have for wind power is in remote areas for raising water from wells for people and cattle to drink. In places where there is no easily obtainable electricity but there is a steady and reliable wind, the modern windmill can be used to turn a dynamo which will make electricity. An example of a modern windmill has been erected in the Orkney Islands. The sails on this windmill, each 5·2 metres (17 ft) long are more like the propellers on an aircraft engine, and the body of the mill is made of steel girders.

But even though wind power is now no longer commercially important, man will continue to use the power of the wind for such pleasurable purposes as sailing, land-yachting, gliding and kite-flying, to mention but a few.

Discovery Assignment 3

1 A part of this chapter mentions the flight to the moon and the problems confronting the astronauts concerning the 'flying' of the stars and stripes flag. How else would the lack of wind affect your life on the moon?
2 Make a weather chart for your classroom but restrict it to facts about the wind, the direction from which it is blowing and its speed. Perhaps you have a device for measuring wind speed, an anemometer, in school. If you have not, you can make a simple one as shown on p. 28. Take wind speed and direction readings at least three times a day.
3 If you were a miller and your windmill was located where you have taken the readings for (2) would it be a successful site for the mill? The information from your weather chart will help you decide this.
4 Where was, or is, your nearest windmill? If it still exists explore it as

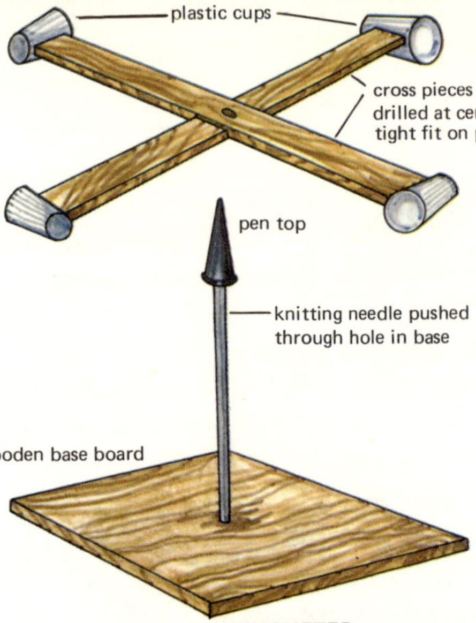

A SIMPLE ANEMOMETER

3:16 (*above*) Unable to find any more thermals (rising columns of air) this glider comes in to land.

3:17 (*right*) As the wind rotates the cups the rotations can be counted during a fixed period.

carefully as you can and make sketches of it. You may have to get permission to do this.

5 With the help of old maps, perhaps loaned from your local museum, or by making a close study of Ordnance Survey maps, locate all the mills in your area. What type of mills were they, and can you date them?

6 If you have a local pond or your own school swimming-pool, make models of sailing ships rigged with different types and amounts of sail, and see if you can discover the most efficient types. Remember that playing near water can be dangerous, so get permission and adult supervision before you experiment.

7 Find out all you can, with pictures and information, about tornadoes. How are they created, where are they most frequently found and how fast do they travel? What usually follows in the wake of a tornado as far as man is concerned? How does it affect his life, and how is his life controlled by tornadoes?

8 Write an imaginative story about (*a*) life in a town threatened by a tornado, or (*b*) being the pilot of an aircraft which is unavoidably having to fly through the 'eye' of a tornado.

9 Are there any modern wind-pumps near your school? If so ask permission to investigate them. Draw or photograph them so that you can study their construction. Take measurements at the base, find the height using a clinometer, and decide for yourself what job they do.

4-WATER

In the last chapter we found out about the story of wind power and how at the present time it has no large-scale usefulness to man. The development of water power began before the power of the wind was harnessed.

Man has always lived, if not beside, within easy travelling distance of a fresh-water supply. His life has always depended on it, as too have the lives of the many different creatures he has kept and still keeps to feed him or to help with his work. Because of man's proximity to water, he soon realised that it was possible to make use of it as a source of power. He only had to wade into a running stream or river to feel the force of the water pushing against his body, and in times of flood, or in larger rivers, the current would sweep him off his feet.

4:1 (*left*) The water wheel at Laxey in the Isle of Man is the largest ever built in the British Isles (see page 33).

4:2, 4:3 Two views of the same cast-iron water wheel, the first showing the cups which receive the water and rotate the wheel, the second showing the sluice-gate mechanism controlling the flow of water on to the wheel.

The first water mill

No one knows for sure who first put a wheel into running water, observed the wheel turning with the current and thus discovered the water mill. Some people give this honour to the Egyptians, some to the Greeks and others to the Babylonians, but whoever it was, we know that they used water mills before the birth of Christ.

The water wheel

The water wheel was brought to Great Britain by the Romans when they invaded and conquered us in 55 B.C., and by this time the wheel had been designed to work vertically in the water, like a car wheel on the road. The earliest type of water wheel had been put to work, very inefficiently, in a horizontal position, flat in the water. The earliest water wheel was made almost entirely of wood, and it looked like any other wheel except that it was broader and the spokes seemed to stick out beyond the rim of the wheel. These paddles, or vanes, were very important as it was against these that the force of the water would be allowed to rush. Attached to the centre, or hub, of the wheel was the drive shaft, usually a heavy piece of timber at least 45 centimetres (18 in.) square. This axle, or drive shaft, passed from the wheel into the mill where it was connected to the millstone runner by means of gear wheels in a very similar way to the windmills described in the last chapter. In the earlier water mills the wheel shaft and gears were all made of wood.

The siting of a water wheel

Obviously if the water wheel had simply been erected anywhere on a river or stream, it would not have been very efficient, so the miller sited his mill very carefully. He wanted a fast-flowing, plentiful supply of water on a river which was not very much wider than his wheel. If he had located his wheel in a wide river, most of the force of the water would flow past his wheel without using the full force of the current to turn it.

Controlling the flow of water

These ideal conditions for siting a water wheel rarely occurred naturally, and the miller was forced to change the river or stream to suit his needs. Frequently he would dam the river above his

TYPES OF WATER WHEELS

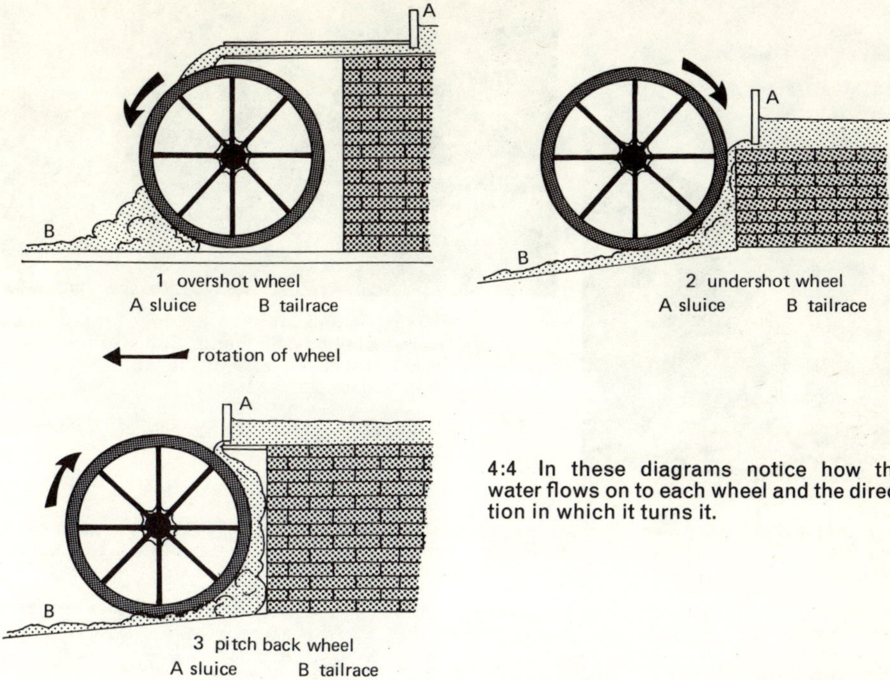

4:4 In these diagrams notice how the water flows on to each wheel and the direction in which it turns it.

mill so that he was sure of having a permanent supply of water. The dam would hold back the force of the stream running down towards his mill and make it possible for him to turn 'the current' on and off, much like we do a tap in our homes today. The miller would not, of course, use a tap; he had sluices to open in order to release the water from behind the dam and let it rush on to his wheel when he was ready to work.

When work was completed, he would close the sluices again and shut off the water supply. He could not stop all the water flowing in the river so he always had to design and build channels for surplus water to continue its flow downstream.

Undershot wheels

The simplest method of getting the wheel to rotate was to place it in such a way that the vanes were struck by the current of water. The water tried to wash the vanes downstream with it

4:5 This natural fountain is one of the famous hot-water springs of Iceland. Can you find out if there are any more parts of the world where hot springs or geysers are found?

4:6 A long-disused tide mill at Trematon, Cornwall.

4:7 A modern tidal power station on the estuary of the River Rance in France makes use of the great difference between high and low water – the moving of the tides – to produce electricity.

and so rotated the wheel and as one vane emerged from the downstream side of the wheel, so another would be submerged on the upstream side, and the rotation continued. This type of wheel was the earliest to be used and was also the simplest form of the undershot wheel. It was called 'undershot' because the water flowed under the axle of the wheel.

As with all things, this undershot wheel was improved, not by altering the wheel but by controlling the water serving it. Instead of allowing the water to flow along a flat bed on to the wheel, a step was built immediately upstream of the wheel so that the water poured on to the vanes higher up nearer the level of the axle. This meant that the water was working its power on the vanes for a longer time as it flowed down and under the wheel thus making the whole turning force more efficient.

Overshot wheels

Each method of supplying the water to the wheel described above turned the wheel in the same direction, but there is one type which rotates the wheel in the opposite way. This is called the overshot wheel because the water is made to shoot, or pass, over the top of the wheel beyond the axle point. To get the water to do this the miller had to design and fix a trough from the river up and over his mill wheel so that at the end of the trough the water fell on to the vanes at the downstream side of the wheel. The overshot wheel was most usually found in hilly or mountainous areas where the streams were narrow but very fast flowing. They were the most efficient of all types of wheels.

Water wheels varied a great deal in size; the largest one ever to be built in the British Isles was 22·1 metres (72 ft 6 in.) in diameter, and it still turns, during the summer months, for visitors to Laxey on the Isle of Man. It is made of cast-iron and was used to pump water from a lead mine 183 metres (600 ft) below the surface. It was made in 1854.

Tide mills

One form of water mill which is quite fascinating is the tide mill. As its name suggests, it made use of the rise and fall of the tide to power the water mill. In some respects this might be called 'moon power' because the moon causes the movement of the tides. As the moon orbits round the earth the tides ebb and

4:8 This water mill straddling a cut beside the Thames is in the process of renovation.

flow twice during a period of 24 hours 50 minutes while the earth rotates once on its axis in 24 hours. This means that the tides are not rising and falling at the same time daily but about 50 minutes later each day. This was the biggest nuisance to the millers working in tide mills. When the tide was rising, the miller had to be ready, no matter what time of day or night, to open the sluices to allow the sea water to fill his mill pond. At the peak of the tide he had to close the sluices to store the water. When he was ready to begin grinding the corn, he would open the sluice to his water wheel and continue work until all the water had drained out of his mill pond. Before he could do any more work, he had to wait for the next rising tide to refill his pond. This meant that the hours he worked were very irregular, but in coastal areas, before any other form of power was available, the work had to be done.

Uses of the water mill

The earliest use of the water mill was for grinding corn but, as we have discovered, it was used for other purposes too. These included the early method of raising water to irrigate crops. Later, when the mills became more efficient and powerful, water wheels turned blades in sawmills, worked tilt hammers in the iron industry and hammers for ore crushing and beat cloth in woollen mills. They were also used in other industries such as copper, paper, snuff and furniture-making.

Steam power

When steam power became available and proved more efficient than water power, the water mills of Britain fell into disuse, just as the windmills did. But the story does not end there for water power. In modern times man, with more knowledge and better materials, has turned back to water and once again made use of this tremendous and never-ending source of power.

Hydro-electricity

The modern use of water power is to produce electricity. When electricity is made in this way it is called hydro-electricity, the word 'hydro' coming from the Greek word *hudor* meaning water.

As the miller did in days gone by, so the present-day engineer must position his hydro-electric plant very carefully, taking many things into consideration. First of all, he must be sure of a plentiful supply of water, which usually means that the rainfall in the area must be high or the rivers serving his plant must collect water from a high rainfall area. He must be certain that the rock in the place where he is going to site the work will hold the water on the surface and not soak it all up. Also he must find an area where the water can have a large drop or fall because water which falls from a great height is many times more powerful than water which flows in a river bed. Finally, he must be sure that he can control the water so that it works for and not against him.

Remote areas

All these problems can usually only be answered in high mountainous regions of the world, but it is frequently these places which are the least inhabited parts. This is important for two reasons, the first being that the place selected may be difficult to get to with the huge pieces of equipment needed, and there may not be local towns from which to obtain the hundreds of workers needed to build these plants. The second reason is that when the plant is built and the electricity is being produced, it must be sent to large areas of population to be used, and this usually means that high-tension power cables have to be erected.

Norwegian hydro-electricity

As you will realise all this is a very costly business and so the siting of a hydro-electric plant has got to be carefully planned and thought out before any work commences.

On a physical map in your atlas, Norway would appear to be an ideal country for the production of hydro-electricity, and it might be thought that the people of Norway would require no other means of producing their electricity. They live in a very mountainous country with a high rainfall, many lakes and fast-flowing rivers falling rapidly into the deep fiords of the North Sea. But the country is so difficult to travel across and work in that although the Norwegians do have many hydro-electric schemes providing them with power, they still rely on other methods as well. Other types of power stations can be built more easily and cheaply where the people are who need the electricity.

4:9 The Stwlan dam in the Ffestiniog scheme in Wales has been built across the river holding back the water to form a lake. This water can be controlled to flow over the turbines to produce hydro-electricity.

4:10 This photograph shows the outer wall of the dam. Notice how strongly it has been made with buttresses to hold back the weight of water in the reservoir.

Building the dam

When the decision has been made to build a hydro-electric station, it generally follows that a huge dam must be built to hold back the water in a lake so that (*a*) there is always water available, (*b*) it will have a powerful fall, (*c*) it can be controlled to pass through sluices on to the 'wheels' which help to make the electricity, and the river below the dam must be allowed to continue to flow uninterrupted.

Great care must be taken when building the dam to ensure that it will always be safe. The rock on which the dam rests must be strong enough to support millions of tons of reinforced concrete. If it is not strong, it has to be excavated until safe rock is found. Very often when this dam building is in progress, the river's course is diverted so that the work can be done more easily. The dam itself has to be designed and tested thoroughly, in the form of models, to make sure it will withstand far more water pressure than it is ever likely to need to do when it is completed. Usually you will find that these dams are built in a curve, as this has proved to be much stronger than a straight dam wall between two banks.

When the work has been completed, the river will be returned to its natural course and a lake will form on the upstream side of the dam. When the water in this man-made lake has reached its designed working depth, the plant can begin its production of electricity.

The turbine

The sluices in the dam wall are opened at the headgate and water tumbles down tubes called penstocks and into the turbine chamber. The turbine is the equivalent to the miller's water wheel, but the miller would certainly not recognise it. There are a number of different types depending on where they are going to be used and what pressure of water is supplied to them. Some are used in a vertical position, like the water wheel in the mill, but the Pelton wheel is the one most like the miller's, having huge cups around its circumference. The Kaplan turbine, on the other hand, could be mistaken for a ship's propeller and is mounted perpendicularly to the flow of water.

4:11 The Pelton wheel – a type of turbine used in hydro-electric schemes.

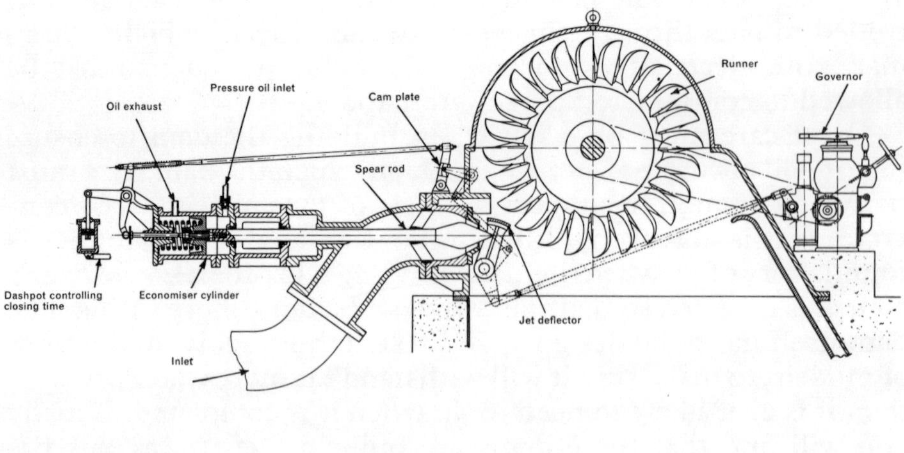

As the turbines are forced round at terrific speeds by the power of the water, they turn shafts which are connected to electric generators. The generators are thus powerfully rotated, thousands of times every minute, to produce electricity. The electricity leaves the hydro-electric plant by cable. How it joins the National Grid and is controlled will be discovered in Chapter 7.

Tidal power stations

It was not until very recently that it was found possible to make modern use of the old-fashioned tide-mill idea. But now in Brittany, in France, where the River Rance joins the sea, there is a tidal power station which began operating in 1966. Sites for tidal power stations have to be carefully selected, like their counterparts in the mountains, because they are only really efficient and productive where there is a large difference between high and low tides. The tidal movement at the mouth of the River Rance is 13·1 metres (43 ft), and the principle on which

4:12 The power from the mighty Niagara Falls which separate the United States and Canada does not go to waste, but is used to produce hydro-electricity.

the hydro-electric scheme works is strikingly similar to the method employed by the miller who owned the tide mill. The tide is allowed to rise to its maximum where it is captured and then released through the turbines to turn the generators to make electricity. The turbines fitted in the River Rance scheme are reversible so that they can help to produce electricity as the tide rises as well as when it falls.

The River Rance hydro-electric scheme, along with the other mountain-sited schemes, produce for France over half of the country's electrical power. Other countries of the world which are using water power to produce electricity include Britain, Norway, Sweden, Switzerland, Germany, Russia and the United States.

Natural waterfalls

In Canada and Central Africa the power of the massive waterfalls of Niagara and Victoria is harnessed to produce electricity.

Hot springs

New Zealand and Iceland are two areas of the world famed for their geysers and hot springs. These are water courses rising from under the ground where they have been heated by the earth's natural heat. When they rise to the surface, they send up steam and boiling water which has been harnessed to provide power and heating in these countries for houses and factories.

Leisure sports

Many of the artificial lakes made by the hydro-electric dams have provided other uses for man. They have become very popular for water sports such as boating, fishing and water-skiing.

Discovery Assignment 4

1 Where is or was your nearest water mill? Again, you may have to do some detective work using old maps and books. If you can locate the site of a water mill, investigate the area closely to discover where the wheel, mill, miller's house and the stored 'head of water' would have been.

2 Make your own model of a water wheel and place it, at varying depths, in running water to see in which position it works most effectively. Do

4:13 Reservoirs form open expanses of water inland which can be used for sports like sailing. Can you think of other leisure activities on reservoirs?

not forget to try it horizontally as well as vertically. Use a stopwatch to count the revolutions per minute.

3 If you have a convenient local stream, make a small boat and sail it downstream, carried only by the current, for a fixed distance. Make a note of the time it takes. Make a temporary dam across the stream at the head of the experimental section; when the water has built up behind the dam, break it and release your boat on the flood of water. Is there any difference in the time taken, and what other changes do you notice?

4 If you have a model steam engine at school attach the water wheel which you used in (2) to it and use the engine to turn the water wheel. Time the revolutions per minute and compare the result with the ones you obtained from the earlier experiment.

5 On a map of the world show the countries which make use of hydro-electricity. To find this information you will have to refer to your geography books and encyclopedias. What do all these countries have in common?

6 Look up on a map of France the River Rance area where the tidal power station has been built.

7 Make a wall display with pictures, writing and models showing how wind and water play very important parts in the leisure periods of our lives.

5-COAL

How it was formed

The story of coal is a long one beginning approximately 300 million years ago. There were no human beings on the earth at that time, and the black rock-like material which we now call coal was, in fact, growing vegetation in the form of plants, mosses and trees. As these died and rotted, they were covered with fresh growth and later with volcanic lava and rock in huge earth movements. The vegetation and other deposits were gradually packed tighter and tighter and compressed to such a degree that now, millions of years later, we extract them in the form of coal, often from hundreds of feet below the present surface of the earth. This 'rock' grew on the surface of the earth in a period of the world's pre-history called the Carboniferous period.

The history of coal

The process of vegetation growing, dying, rotting and being overlaid with rock happened a number of times during the millions of years covered by the Carboniferous period, and in this way it is now possible for us to find many different seams of coal beneath our feet, some quite close to the surface and very thin, others very deep down and much thicker.

There is very little doubt that the ancient Britons knew about the black rock which burned brightly and gave warmth, and we know that it was being mined in Britain at the time of the Roman conquest of these islands. Coal stores have been discovered in various places along the length of Hadrian's Wall, a Roman fortification in the extreme north of England.

It was at this stage in our history that coal, as a source of power, began to take over from the water and wind power of earlier times. Moreover, with the invention of steam vehicles on rails, this form of power began to bring about the end of the importance of canals as a major method of cheap transport for bulk materials. The new railways were easier to build, faster and more convenient, because they carried passengers as well as goods.

The history of coal mining, including drift mining and bell pits, and the continued development and mechanisation of our coal mines through to the present day, can be studied in the

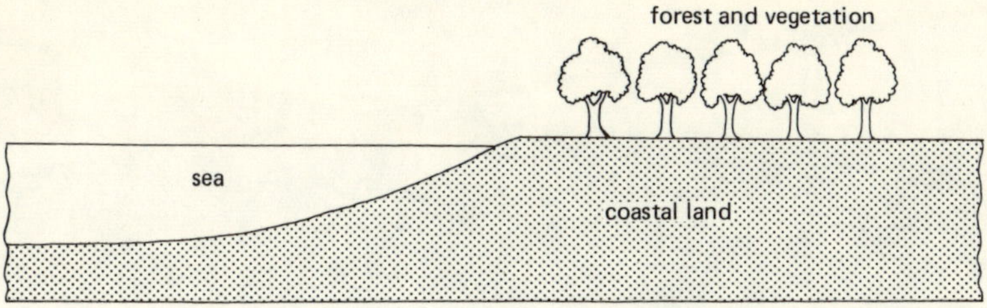

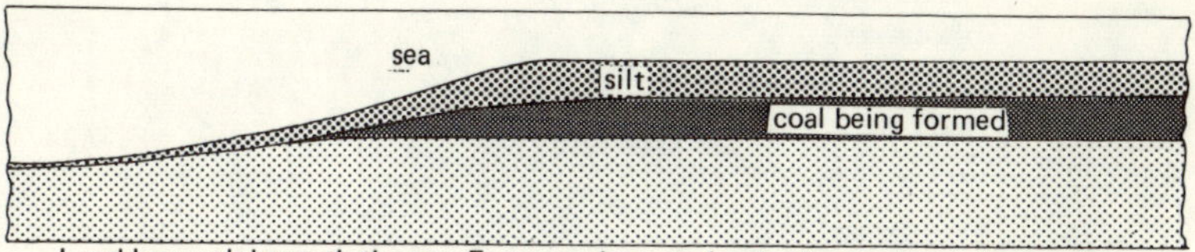

Land has sunk beneath the sea. Forest and vegetation begins to change into coal.

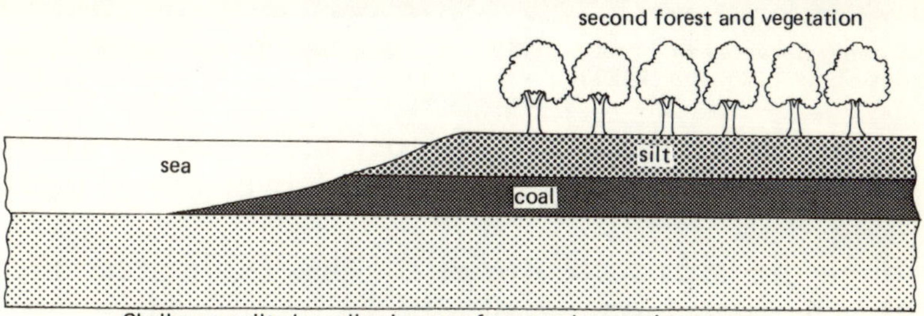

Shallow sea silted up allowing new forest and vegetation to grow.
Coal-forming process continues below ground.

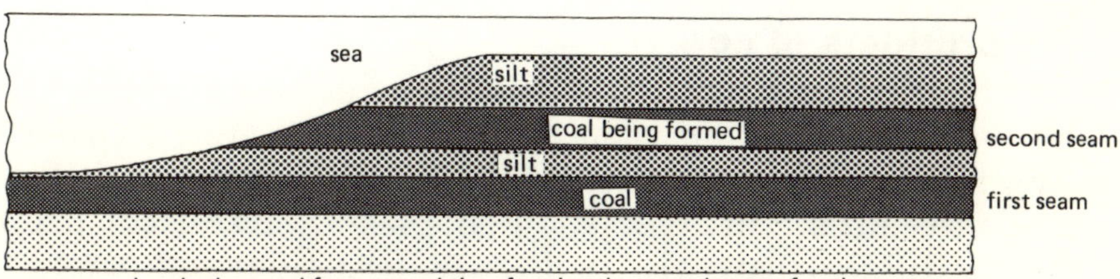

Land submerged for a second time forming the second seam of coal.

5:1 When studying these diagrams showing how coal was formed remember that there were hundreds of millions of years between the first stage and the last.

5:2 Coal being delivered to a power station on a loop line from the main rail system. The train carries about 1,000 tons (1,120 tonnes) of coal.

Environmental Studies Book Number 11, *Minerals, Mines and Mining*. Chapters 4 and 5 of this book will prove especially interesting.

Through the last two centuries coal has had a great effect on the industrial development of this country, and this can be traced through improved methods of transportation, first with canals, later with railways, and by using stationary steam engines to power factories and pump mines dry of water.

By-products of coal

Although we are interested in coal as a source of power, it is fascinating to find out what other uses there are for coal and its by-products. Here are just a few of the everyday things which are made from coal: detergents, nylon, plastics, fertilisers, printing-ink, perfume, chemicals, explosives, pesticides and paint.

From coal to electricity

About half the coal mined in Britain goes straight to power stations to be burned as part of the process of making electricity.

As it is easier and cheaper to move electricity about the country along cables than it is to move coal, most of the new power stations using coal are built near the coal-fields. Others are located close to the railways near the ports. Where coal has to be transported, special equipment is used. By sea, coal ships, or colliers, are used; by rail, new designs of special hopper wagons run continuously. Frequently they form a loop from the main railway to the power station, where they shed their load of coal into bunkers beneath the rails and continue their journey to complete the loop back to the main line for another load.

Inside a power station

Large stocks of coal are kept at the coal store in case of emergency, such as the miners' strike in the winter of 1974. The coal for the power station is moved from the store along a conveyor belt which

5:3 This collier is having its hold loaded with coal.

5:4 Here we can see the coal going down the chute into the ship's hold.

eventually feeds into a bunker. From here the coal falls into a pulverising mill which grinds it to a very fine powder. The mill is rather like a large, round metal table on which are placed steel rollers or balls. As the table revolves, the coal is forced under the rollers and crushed.

How electricity is produced from coal

Meanwhile, air is sucked from the top of the boiler house by means of a fan and passed through a section which heats the air. It then moves on to be separated so that some hot air goes directly to the burners of the furnace while the rest, propelled by a fan, passes through tubes to the pulverising mill. Here it mixes with the powdered coal forcing it along more pipes to rise again with the other air in the furnace burners. This produces a very high temperature and is a very efficient method of burning coal.

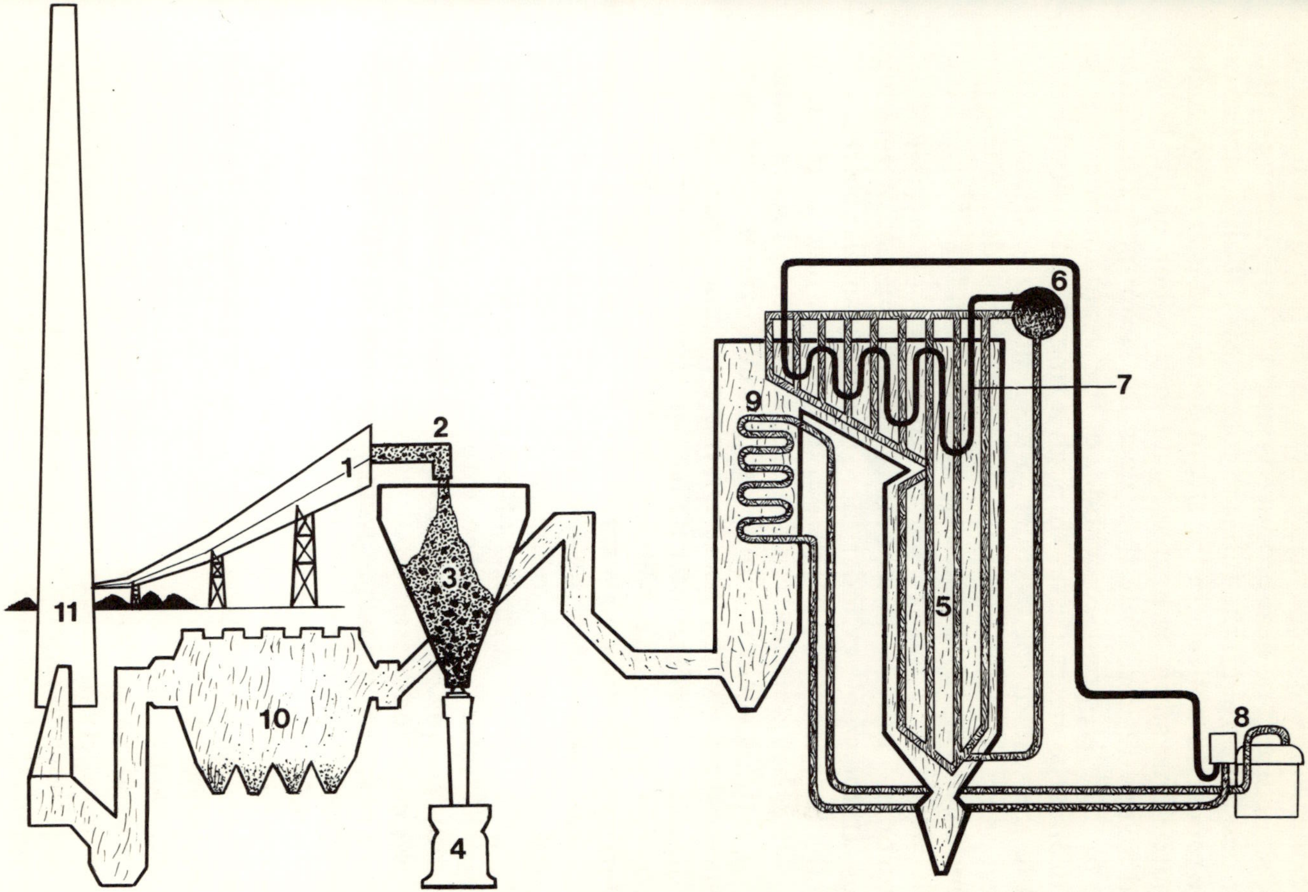

5:5 This very simplified diagram shows the major parts of a coal-powered electricity generating station. 1 conveyor belt, 2 coal tipper, 3 bunker, 4 pulverising mill, 5 furnace, 6 steam drum, 7 superheater, 8 steam turbine, 9 boiler reheater, 10 filter for solid matter, 11 chimney.

The steam turbine

The boiler is made up of many tubes extending the full height of the structure. The heat produced raises the temperature of the water passing through the tubes to form steam. This steam rises to the top of the boiler into the steam drum and is at a very high pressure. This steam is heated again in a section of the boiler called a superheater. Only then does the steam pass through a valve to the high-pressure cylinder of the steam turbine. Inside the turbine the steam passes through a ring of stationary blades fixed to the wall of the cylinder. These direct the steam at the correct angle on to the second ring of turbine blades. These are moving blades attached to the turbine shaft, and the force of the steam causes them and the shaft to rotate. This is one 'stage' of the turbine – one set of stationary blades, one set of rotating blades. In the whole cylinder there are many similar stages.

By the time the steam has passed through the final stage, much of its energy has gone, and it is returned to the boiler for reheating. It then returns again to supply its power to the intermediate-pressure cylinder where there are more turbine stages for it to pass through. From the intermediate-pressure cylinder it passes directly to low-pressure cylinders. In these the steam flows outward in opposite directions through the turbine. With the temperature and pressure of the steam falling it is of no further use, so it is allowed to pass into the condenser which contains about 20,000 pipes with a diameter of 2·5 centimetres (1 in.) through which cold water from a river, estuary, sea or cooling tower is constantly passing. When the cooling steam comes into contact with the cold condenser pipes, it condenses (turns back into water). From here the water eventually finds its way back to the boiler where the whole process can begin again.

The whole of the operation just described is, of course, continuous when the power station is working to produce electricity. There is always superheated steam in the high-pressure cylinder and reheated steam in the intermediate- and low-pressure cylinders, so that all the turbines are rotating the main generator shaft to produce electricity. This shaft usually rotates at 3,000 revolutions every minute (3,000 r.p.m.) and to maintain this the conveyors have to supply 200 tons of coal to the power station every hour. In the meantime, the condenser has to work just as hard as any other part of the plant because it is using more than

5:6 These gigantic structures are called cooling towers and they are used to cool down the water which has been used to condense the steam in the power station.

5:7 Steam power stations are built close to a plentiful supply of water and this photograph shows the water intake point from the River Thames to supply Didcot power station.

225 million litres (50 million gallons) of water every hour to cool and condense the steam back into water. While the cooling water is doing this job, it becomes heated itself and when it has passed through the condenser it is pumped up into huge towers where it is allowed to cool before being returned to the condenser. It is easy to pick out a power station when you are travelling about the country because its curved-sided, chimney-like structures, which are the cooling towers, are huge and, therefore, are visible from some distance, and in a large power station there may be six or more of them.

The other tall structure seen amongst the cooling towers of a power station is the chimney which is connected to the furnace. The remains of the coal and gases leave the furnace and are taken towards the foot of the chimney. On their journey they are treated, so that when they leave the top of the chimney they cannot pollute the atmosphere or the land around the power station. The unburnt ash, for example, is not allowed to pass beyond a set of special plates which attract all the ash to them, allowing the gases to continue their journey to the chimney. This ash, dust and grit is knocked off the plates from time to time so that it falls down into hoppers below. These hoppers are set above the

5:8 This photograph of a power station shows many important features. On the right is the coal conveyor belt into the main turbine house. Behind this we can see the cooling towers and chimney, and in the foreground the pylons and cables carrying the electricity to homes and factories.

ground so that a lorry can pass underneath them to load the ash and drive off to dispose of it in disused pits or sell it for use in road and building construction.

Other large consumers of coal are: the iron and steel industries for use in their coke ovens; factories, hospitals and offices which still use coal for heating purposes; and domestic heating units in our homes.

5:9 Coal for household use can now be delivered in this automatic bagging vehicle.

Gas from coal

Another user of coal is the British Gas Council, but its interest in coal is becoming less important now as we shall read in the next chapter. Britain's gas industry can really be said to have started about 200 years ago when a man called William Murdoch (1754–1839) saw spurts of flame jetting out from coal burning on his fire. He decided to experiment, and put some coal in a kettle and heated it over a fire. He waited until the coal in the kettle was very hot and then put a lighted taper to the spout of the kettle. He found that even when he removed the taper the flame continued to burn at the spout.

This gave Murdoch the idea of lighting houses, workshops and even streets by gaslight. The first gas-lit streets in the world were in London in 1808. Soon afterwards many streets in other towns were lit by gas, and the gas era had begun. It did not take long for people to realise that gas gave off heat as well as light, and by the middle of the century gas cookers and fires were being produced. With the nation using more and more of this new power from coal, gas works were built in many towns.

5:10, 5:11 One of the early forms of street lighting was the gas lamp (5:10, *left*). Gas holders (5:11) have been a feature of our towns for many years. Originally they were used to store town gas; more recently many of them have been converted to store natural gas.

Town gas

To produce gas, the coal has to be baked in large ovens called retorts. This baking in the retorts softens the coal, which gives off a thick, brown smoke. This smoke has to be cleaned because it contains tar, sulphur and ammonia. When this has been done the resulting 'town gas' is stored in tall, cylindrical gas holders ready for use.

By-products

A number of useful by-products now remain. The first of these is coke, the gasless coal, which is used for heating boilers in factories and some schools. The tar, sulphur and ammonia can be used for making many things such as fertilisers, medicine, plastic and nylon.

Although 'town gas' is becoming less important, there are still 1,000 town gas holders in towns in Britain. They had to be built because, although a steady supply of gas was made, people tend to want to use it in large quantities at regular times each day. Therefore, a reserve quantity had to be held to maintain a steady supply to all users at all times. If you have a gas holder near you, observe it closely at different times during the day. You will notice that it seems to be taller at certain times than it is at others. The taller it is, the more full of gas it is. The holders are made to expand upwards like a telescope as the gas is pumped in at the bottom from the gas works.

Discovery Assignment 5

1 A visit to your local coal merchant will probably give you the opportunity of obtaining samples of many different types of coal. He may be able to give you information about each type, but if not, he will certainly supply you with their names, and you can research to find out where they were mined or made.

2 The samples from (1) can be used to start a coal display in your school (do not forget to include peat and wood). From this beginning you can add your written discoveries about the history of the coal, where and how it was mined and what its present-day uses are. Models, made by you, would be very useful in this display.

3 Your coal merchant will also be able to give you the address of your region's Coal Board offices, and if you write to them they will be able to supply you with information about the coal-mining areas of Britain, which you can locate on a map. Do you notice anything about the distribution of coal mines in Britain?

4 Where is your nearest power station? Is it a coastal one or is it situated near the coal mine? Perhaps it is not a coal-powered station at all. If it is not, how is it powered?

5 Do you have gas in your town or village? If you do, what type is it? If it is town gas, you will probably know where the gas is made and where the gas holders are. Keep a check on the gas holders to see when they are full and when they are empty. Does this occur at regular times each day? If so, when?

6 If you are on town gas try and find out when you are going to be converted to natural gas, and how the Gas Council organises this very complicated change-over.

7 If your area has recently changed to natural gas, where were the old gas holders, what has happened to them and what is the land now used for?

5:12, 5:13 Gas cookers have certainly changed considerably since the one on the left was in use in the 1890s. The one on the right dates from 1923.

6-GAS AND OIL

Town gas, as we have just discovered, is not a power source in its own right. It is really the coal which provides the power. Gas has also been produced by burning oil instead of coal, but both these processes destroy one important source of power to produce a secondary one. Nevertheless, for 150 years town gas has been a useful power supply for homes and industry in this country. At the same time, however, it has created in its production a great deal of pollution in our towns and cities. Fortunately, we are having to rely less and less on this type of manufactured gas, and the Gas Council forecast that soon no town gas will need to be made.

Natural gas

What is now replacing town gas and will do so even more in the future? Many of us know already because we receive supplies of it – the answer lies in natural gas. It is called natural gas because it does not have to be made; another power supply is not destroyed to enable us to make use of it. Like some other sources of power, it is to be found deep underground in certain areas of the world. Because the creation of natural gas, over a period covering millions of years, is somewhat similar to that of oil, and they are sometimes found in the same areas of the world, this chapter will tell you the story of both natural gas and oil.

THE FORMATION OF GAS AND OIL

The processes starting the long chain of events which gives us oil and gas to use could have started 400 million years ago. It all happened in the sea, which at the time was teeming with fish, microscopic animals and plants. As these died and settled to the bottom of the oceans they would, as time passed, be covered with silt and sand, possibly brought down by the great rivers, and with rock washed down from the land masses. Gradually, these layers were forced deeper and deeper and compressed under the weight of material above. This caused very important changes to take place and, during the course of millions of years, tiny droplets of oil and gas were formed.

Earth movement

Even today we are aware of tremendous earth movements in the

THE THREE MOST COMMON TRAPS WHERE OIL AND GAS CAN BE FOUND

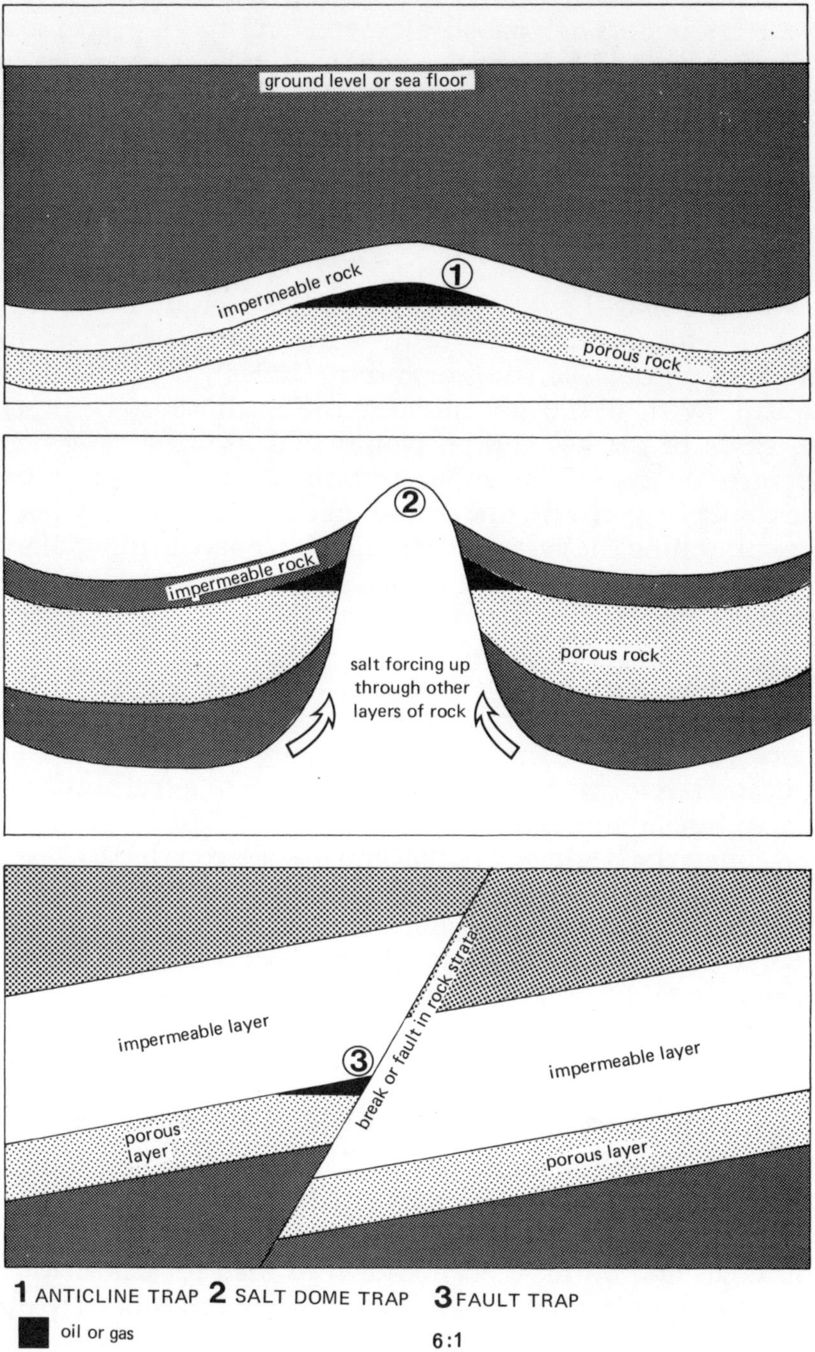

6:1

form of earthquakes, tremors and volcanoes and yet this period in the earth's history is probably quieter and more peaceful than it has ever been. Think how much earth movement there must have been millions of years ago with layers of rock, strata as they are called, sliding slowly but surely on to each other in the earth's crust, sometimes grinding and tearing, sometimes buckling and lifting in great wave-like curves.

Natural reservoirs

This meant that the oil and gas droplets would also be on the move. Their movement might be sideways or, if the rock above them was porous (i.e., full of minute holes), they could travel upwards. With all the movement in the earth it is to be expected that much of the gas and oil would find itself forming natural underground reservoirs either combined or as separate fields. This would happen on a line of fault in the earth's strata where the layer containing the gas or oil was no longer able to move along its porous strata because the fault had brought it up against a non-porous, or impervious, layer.

Anticlines

Another common natural reservoir can occur in a geological feature called an anticline. This is really like an upside-down basin, and it is formed by pressures in the earth's crust causing the strata to buckle upwards.

Although the beginnings of oil and gas were in the sea, it is now quite likely that oil-bearing rocks are also under the dry land. The major oil and gas supplying countries of the world prove this as they are the Middle East, the Middle West of the United States of America, and Russia. But there are still supplies of oil and gas found near the seas and oceans where they were formed, as in the Caribbean and North Seas.

EARLY USERS OF NATURAL GAS AND OIL

Most of us think of oil and gas as being very modern sources of power, related to the development of cars, trucks, diesel engines on rails and at sea, and of high-speed jet aircraft, but in fact natural gas and oil have been known to man for centuries. The Chinese were using natural gas to heat brine (sea water) to obtain the salt which they needed. They probably discovered the gas leaking out of the surface of the earth and found that it burnt readily and provided heat. They set to work and formed the first gas pipeline by connecting together hollow bamboo pipes leading from the gas source to the houses where the salt-making process was being used. In Egypt and Persia oil found on the surface was used for heating and lighting, and Marco Polo told of the oil springs at Baku which he visited in the 13th century.

The first 'wells'

The modern development of greater and wider uses led to more thorough and accurate searches for gas and oil. These were started in 1820 by William Hart, an American. He drilled the first gas well near Fredonia in New York state. It was less than 9 metres (30 ft) deep, but the gas obtained was sufficient to pass through lead pipes to provide lighting to a number of houses in the town.

'Wildcatters'

Still in America, but 39 years later, Edwin L. Drake struck oil with his well 18·6 metres (61 ft) deep, near Titusville in Pennsylvania. This discovery sparked off the era of the 'wildcatters' – men who flooded into the area sinking wells without any geological knowledge, in the hope of finding oil and 'striking it rich', thus making their fortunes. Many did, and lost their money again by wasting it on gambling or further less successful wildcatting.

6:2 In this American oil working of 1875 the power for the pump is supplied by a steam engine in the shed on the left.

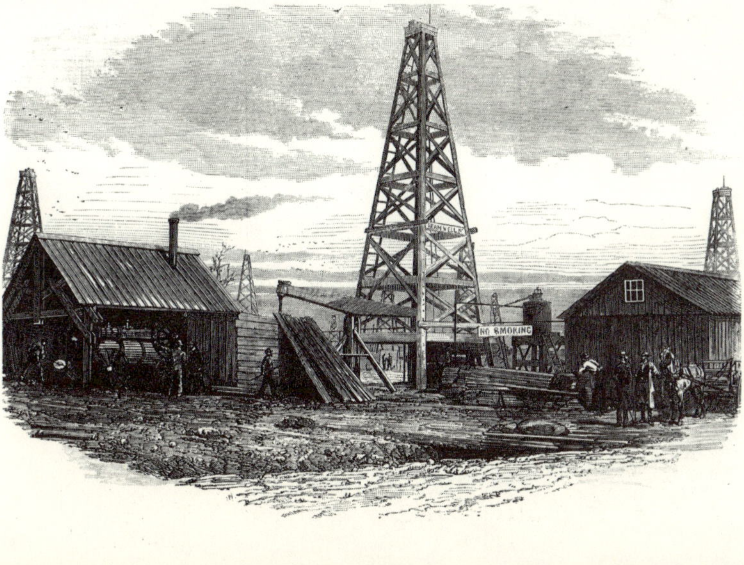

GEOLOGICAL SURVEYS

Nowadays gas and oil is never searched for by guesswork or luck as it was a century ago. Now it is a scientific operation conducted by very clever geologists using expensive and accurate equipment. In a full survey, photographs of a chosen area may be taken from the air to be studied to reveal the true shape of the surface of the land. Geologists will work in the area on foot taking samples of the rock on mountains and river beds, or noting the strata in cliff faces. Most important of all they will arrange for a seismic survey. At sea this type of survey is the best one available to the geologist, for he cannot explore the sea bed or photograph it as he can on dry land.

Seismology

Seismology is the study of earthquakes. In a seismic survey, the geologist makes his own mini-earthquake by creating an explosion

6:3 'Christmas tree' installed at the head of an oil well at Fahud, Oman.

in the sea or about 15 or 30 metres (50 or 100 ft) below the land surface. Before setting off the explosion, he places recording meters at various distances away. These meters, called seismometers, will receive and record the shock waves from the explosion as they bounce back from the different layers of rock strata beneath the sea or land. With these recordings the geologist will be able to put together a fairly accurate picture of the shape of the rocks underground.

By studying all the information he has collected, he will have to make a decision whether it would be worth while drilling a pilot well to find out if gas or oil really lies below the surface where he thinks it does. In the end the only true way to find out is by sinking a well. This is such an expensive business, especially now when most of the well-known oil and gas fields are reaching the end of their usefulness, and the new fields are in far-away or difficult places to reach, that the geologist has to be as accurate, careful and sure in his work as he can be.

6:4 Men working to install a well head.

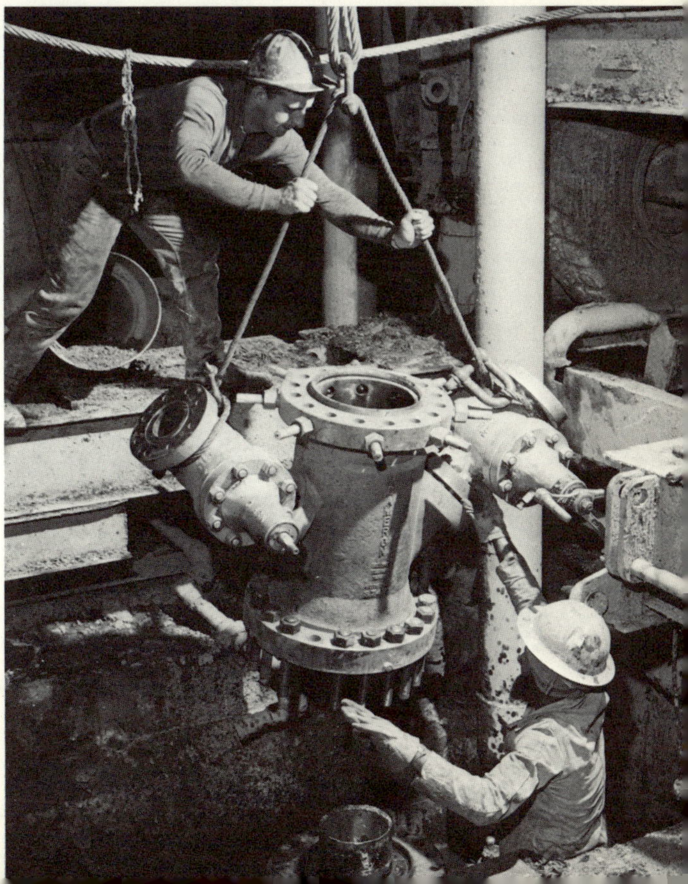

6:5 Section of pipe being made ready for erection at a refinery in Buenos Aires.

Sinking a well

If the pilot well proves successful in finding gas or oil in quantities sufficient to make it worth while spending vast sums of money to bring it to the surface and to transport it where it is required, the oil or gas field is opened and more and bigger wells are sunk. Wherever these 'fields' are discovered – whether they are in the desert, the Arctic or under the sea – very similar drilling techniques are used. Wherever drilling takes place, it is a difficult and dangerous occupation. The men have to work hard for long periods, often in dreadful conditions.

The derrick

The derrick is erected on the site where the well is to be sunk. The derrick is a tall, tapering tower of steel up to about 45 metres (150 ft) tall. From the top of this hangs a rotating joint – the swivel – supported by a block and tackle. The swivel is connected to the drilling string, which is the hollow pipe in nine-metre sections which follows the 'bit' as it cuts its way down through the rock below.

6:6 These drilling bits are used on the oil rigs and consist of three segments, each fitted with a conical cutter.

The bit and the kelly

The bit is the cutting tool for the well; it is large and heavy, with the cutting edge toothed. The only thing required in order to start the work is power to turn the drill-string and hence the bit. This can be provided by electricity, steam or diesel engine and it powers the table, a rotating section of the floor of the derrick. The centre of the table has a square or six-sided hole cut into it which the 'kelly' – a matching length of steel attached to the top of the string – locates to drive the bit round.

Drilling-rig vocabulary

Have you noticed that to work on a drilling rig it seems that you have to learn a new vocabulary? There are other interesting new words to learn concerning the drillers themselves. The driller is the man who leads the crew, a derrick-man climbs the derrick to hitch on new drill-pipes, the pipe-racker is in charge of the pipes, spanner-men look after the rig itself and roustabouts are odd-job men. Roughnecks are new men on the job, learning the trade.

Drilling on land

Work begins when the bit is attached to the drilling string and is lowered on to the earth. The kelly engages in its place in the table and the table begins to rotate, turning the bit which bites into the soft surface rock. Down and down goes the bit until the kelly almost disappears after it. The table is stopped and the kelly is hoisted out until it can be disconnected and a new length of piping inserted. The kelly is re-clamped and lowered, and drilling can commence again.

To help the expensive bit to do its work more efficiently, to keep it cool and to bring up the waste rock, a mixture of mud and chemicals is pumped down the hollow string to the cutting face. The average life of a bit is about 60 metres (200 ft) and in a well over 3·2 kilometres (2 miles) deep, as some of the North Sea gas wells are, this means that many new bits are used.

6:7 This well has been fitted with control units to the pipeline to allow the oil to flow.

A 'gusher'

If everything goes well, after weeks or months of drilling the bit will eventually break through into the underground reservoir of gas or oil. In these later stages precautions are taken at the well head in case the drill strikes a 'gusher' – gas or oil under such pressure that it surges up the well destroying the rig, harming the men and possibly catching fire!

A 'Christmas tree'

When a successful and safe well is drilled, it must be lined with permanent pipes and controlled at the head by a 'Christmas tree'. This is the name given to the unit of pipes and valves which keeps the gas or oil in the well under control until a permanent pipeline overland (or under the sea) is connected between it and the terminal which is going to serve as the distribution point.

Drilling at sea

We have followed through the drilling operations of a rig on land, but frequently the same job has to be done at sea. This is even more hazardous because not only have the men to concern themselves with the drilling, but they have the worry of gales at sea beating on their rigs. In the early days of drilling in the North Sea, thirteen men lost their lives when their rig sank in a gale.

6:8 Three 'roughnecks' on the derrick attaching a new section of drill pipe.

Sea rigs

The four main types of sea rigs are shown in the diagram; they may not be beautiful, but they are certainly built for the job. The Gas Council/Amoco Group rig *Orion* is the mobile platform type, and its legs are about 118 metres (387 ft) long. It can work in 84 metres (275 ft) of water, drill to a depth of 6 kilometres (20,000 ft), withstand waves of up to 19·5 metres (64 ft) high and winds of 184 kilometres per hour (115 m.p.h.).

6:9 This view from the top of the rig shows the 'roughnecks' at work on the derrick floor.

FOUR DIFFERENT DRILLING RIGS AT SEA

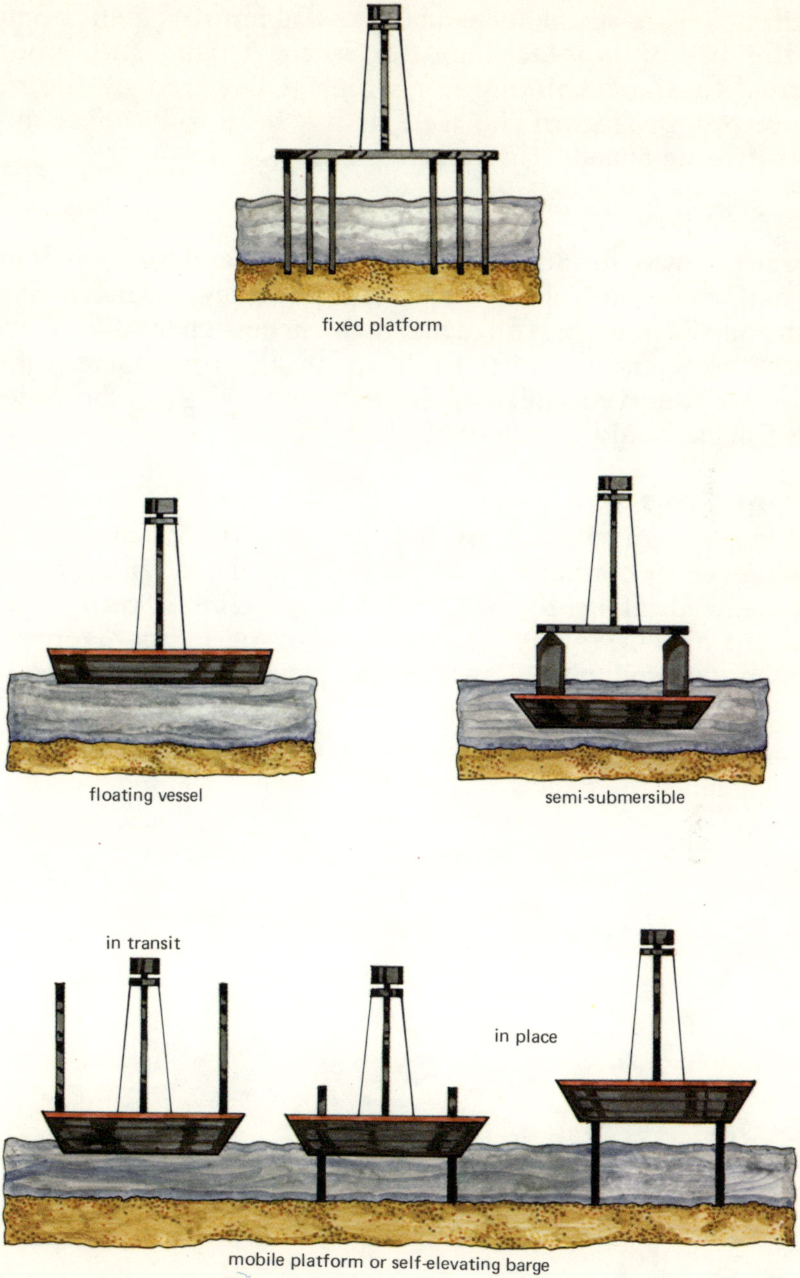

6:10 Notice how some of these four types of rig have their drilling platforms above sea level and some stand on the sea bed while others float on or beneath the surface of the sea.

Turbo-jet drilling

Recently a new drilling method has been developed called 'turbo-jet drilling'. This seems to be a very useful improvement because only the drill bit is rotated and the string remains still. The bit is turned by the turbine motor which is powered by the mud mixture pumped down the well. It has been found that much higher drilling speeds can be achieved.

North Sea gas

Natural gas was found in Holland in 1959 and since that first find, nations on both sides of the North Sea have been drilling to find more. We in Britain have had little success on land so we were forced to search in the North Sea itself, and by 1966 it was realised that we had found enough to indicate that the whole of the nation's needs for gas could be satisfied from this one area.

The pipelines

The big problem to overcome was bringing the gas ashore. This was achieved by connecting a pipeline, from a fixed platform over the well-head, along the sea bed. The pipeline is composed of lengths of pipe up to 76·2 centimetres (30 in.) in diameter, made of steel but coated with glass fibre and pitch to prevent corrosion,

6:11 Lay barge *Hugh W. Gordon* laying gas pipelines on the bed of the North Sea.

and further coated with 8–10 centimetres (3–4 in.) of concrete to make sure that they sink to the sea floor.

Laying the pipeline

All the North Sea pipelines have been laid by an American lay barge, the *Hugh W. Gordon.* Several lengths of the pipe are strung along the side of the barge and their bare metal ends welded together at four welding stations. The welds are checked by X-ray and then wrapped and sealed for protection, thus making a continuous pipeline. As the length of pipe increases, it is passed down a sloping bridge, off the stern of the barge, called the 'stinger'. This passes downwards and behind for 183 metres (600 ft) to support the pipeline down to the sea bed. As each section is lowered into place, the barge moves forward the same length. This process is repeated until the shore is reached.

The jetting barge

Following behind this laying process is a second vessel called a 'jetting' barge, which uses jets of water under high pressure to cut a trench in the mud beneath the pipe. This buries the pipeline between one and three metres below the sea bed where it can rest, safe from the effect of currents and tides.

6:12 British Gas Corporation's third coastal terminal at Theddlethorpe, Lincolnshire.

6:13 Gas holders at the Theddle-thorpe terminal.

6:14 Gas storage containers. In the background is the old-fashioned gas holder and in the foreground is a high-pressure gas container.

At the gas shore terminal

When the pipeline is connected to the shore terminal, the gas needs some processing before it is ready to flow along the gas mains to our houses. It has to be cleaned to extract any water or solids and checked to make sure that it burns evenly and produces the amount of heat that it should. Its pressure is controlled so that the equipment using gas in our homes and industry will work efficiently, and it is also given a smell so that if it leaks anywhere it can be detected and traced. At the present time there are three shore terminals controlling our natural gas from the North Sea.

6:15 Gas pipe location sign.

6:16 Natural gas pipeline marker.

6:17 Pressure and flow control valves in a cross-country natural gas pipeline.

6:16

6:15

6:17

6:18 View of the deep-water harbour at Fawley, where oil tankers unload their cargo at the refinery.

At the oil refinery

Oil straight from the well cannot be used until it has been processed in a refinery. One thing which all refineries have in common is their location close to the sea, often with their own deep-water harbours in which to dock the huge oil tankers so that their cargoes of oil can be discharged. Large quantities of water are also required in the processing of crude oil. The crude oil can range from pale yellow, and be very runny, to dark green or black, which can be difficult to pour. Whatever it is like it has to be distilled, that is, boiled to change it into a vapour and then condensed by cooling it.

The fractionating tower

The different types of crude oil boil at different temperatures and condense at different temperatures. By using this knowledge the refinery can separate light oils from medium and heavy ones. Most of this work is done in the 'fractionating tower' in which the oil is designed to be at different temperatures, cooling down as it reaches the top. At fixed heights up the tower trays are located to receive the oils as they condense. The oils of each grade can therefore be tapped off at different heights.

The uses for different oils

This is the major but not the only process in producing the different oils and fuels which we require to power many of the engines used in our everyday lives. The heavier fuel oils, from the lower part of the fractionating tower, are used by ships and large industrial furnaces; diesel oil, from higher in the tower, is used by lorries, buses, railway engines and tractors; kerosene or paraffin is used for domestic heating and high-speed jet and turbo-jet aircraft; and petrol, from the top of the tower, is used to power cars, piston-engined aircraft and a host of small machines. To keep all these engines running smoothly, efficiently and with a minimum amount of wear to the moving parts, a large variety of lubricating oils are produced.

6:19, 6:20, 6:21 Petrol distribution point at a railway siding. (6:19) Notice the connecting hoses and pipes beside the track. (6:20) Storage tanks. (6:21) Tanker loading bay. Pipelines load from the storage tanks to the top of this hooded section so the tankers can draw up underneath to be filled up.

6:20

6:19

6:21

Exhausting supplies

It has been estimated that at the rate we are using oil from the earth, we only have enough left, of that which is now known to be underground, to last until about 2010. This makes one understand why the search for oil is carried out so eagerly in all parts of the world, and how really urgent it is to find new sources of power very soon.

6:22 This Gas Board converter is altering a gas cooker so that it can be used with natural gas.

6:23 This pipe which is being laid is part of the gas pipe that connects the coastal terminal at Bacton with the main transmission system.

Discovery Assignment 6

1 Make a series of models of offshore drilling rigs like the ones to be found at work now in the North Sea.
2 Collect all the information and photographs you can of present-day earthquakes and volcanoes.
3 Make a study of one of the historical volcanic eruptions, like the one at Pompeii, or the great San Francisco earthquake. Compare the information you have gathered from (2) and (3) and you will realise how mighty is the power held in the earth.
4 You have read about a geological feature called an anticline. You can make an anticline by holding a strip of card at each end, pressed flat on a desk, and slowly pushing your hands towards each other. The card will rise in an anticline.
5 Write an imaginary story of a 'wildcatter' of North America in the 1860s, his search for oil and the making of his fortune.
6 It would be difficult for you to make a seismographic survey of your area, but it is very easy for you to make a surface geological survey by collecting typical samples of soil and rock from a radius around your school. Assemble your collection in compartments in a box and label each sample with the type of rock and the place where it was found.
7 From your rock sample survey you can produce a simple geological map of your area.
8 You might be lucky enough to find an area which is being supplied with natural gas now. Investigate the work, the siting of the pipeline, the trenching, the pipeline itself, the machines doing the work, and the completion of the job when the topsoil is replaced and levelled.

7-NUCLEAR OR ATOMIC ENERGY

Power from atoms is unlike any other power source we have read about so far, and it is really a form of chemical energy. To understand how we get power from atoms we must first understand a little about the chemistry of the world around us. So far in this book we have been 'thinking big' – of thousands of men working together, of winds to turn huge sails on windmills, of the powerful force of the water stored behind great dams, of extensive coal mines and deep oil wells – but in this chapter we must first think small, so small that the things with which we are concerned cannot be seen.

Atoms

Everything around us is composed of tiny atoms of chemical elements. There are nearly one hundred different sorts of atoms. Each one is made up like our solar system where the planets revolve round the sun. The planets in the atom are called electrons, and the sun is called the nucleus. The electrons are held in orbit round the nucleus by a steady electrical force. The nucleus at the centre of the atom is not one solid mass, but is usually made up of neutrons, which have no electrical charge, and protons, which have a positive electrical charge. Because the electrons are negatively charged they cancel out the positive electrical charge from the proton part of the nucleus.

Isotopes

Every element has a different number of protons (and usually a matched number of electrons), and it is this which makes the elements different from each other. Hydrogen is the lightest of all the elements and has only one proton and one electron in its nucleus. At the other end of the scale is uranium, the heaviest natural element with 92 protons. Each element may be found in different forms, and these are called isotopes. Isotopes of the same element all have the same number of protons, but different numbers of neutrons. Uranium is important in atomic power. In its natural state it has three different isotopes, the commonest having 92 protons and 146 neutrons. When naming a uranium isotope the protons and neutrons are added together, $92+146=238$. So this isotope of uranium is called U238. The two other isotopes are U235 (92 protons, 143 neutrons) and U234 (92 protons, 142 neutrons).

7:1 The huge mushroom-shaped cloud was created by the under-sea detonation of an atomic bomb. During this experiment a battleship was carried up by the giant water spout and it plunged back into the ocean and sank.

Nuclear fission

If a neutron is shot at the nucleus of a uranium 235 atom, it will break up into parts making two smaller atoms, and it will release energy, mostly in the form of heat, and two or three neutrons will be given off. These spare neutrons hit other nuclei (more than one nucleus), and the process continues. This process is called 'nuclear fission', and it is a chain reaction; it will keep on working with more and more spare neutrons splitting more and more atoms. The important thing for us in this process is that it produces great heat. The problem confronting the scientists who developed nuclear fission was that having started the chain reaction, how could they control it and make it useful to man as a very important source of power? If it is not controlled, the nuclear fission is so fast and violent that a devastating explosion occurs. It was two such explosions from atomic bombs – uncontrolled nuclear reaction – which ended the war against Japan in August 1945 when the bombs were dropped from American bombers on the cities of Hiroshima and Nagasaki.

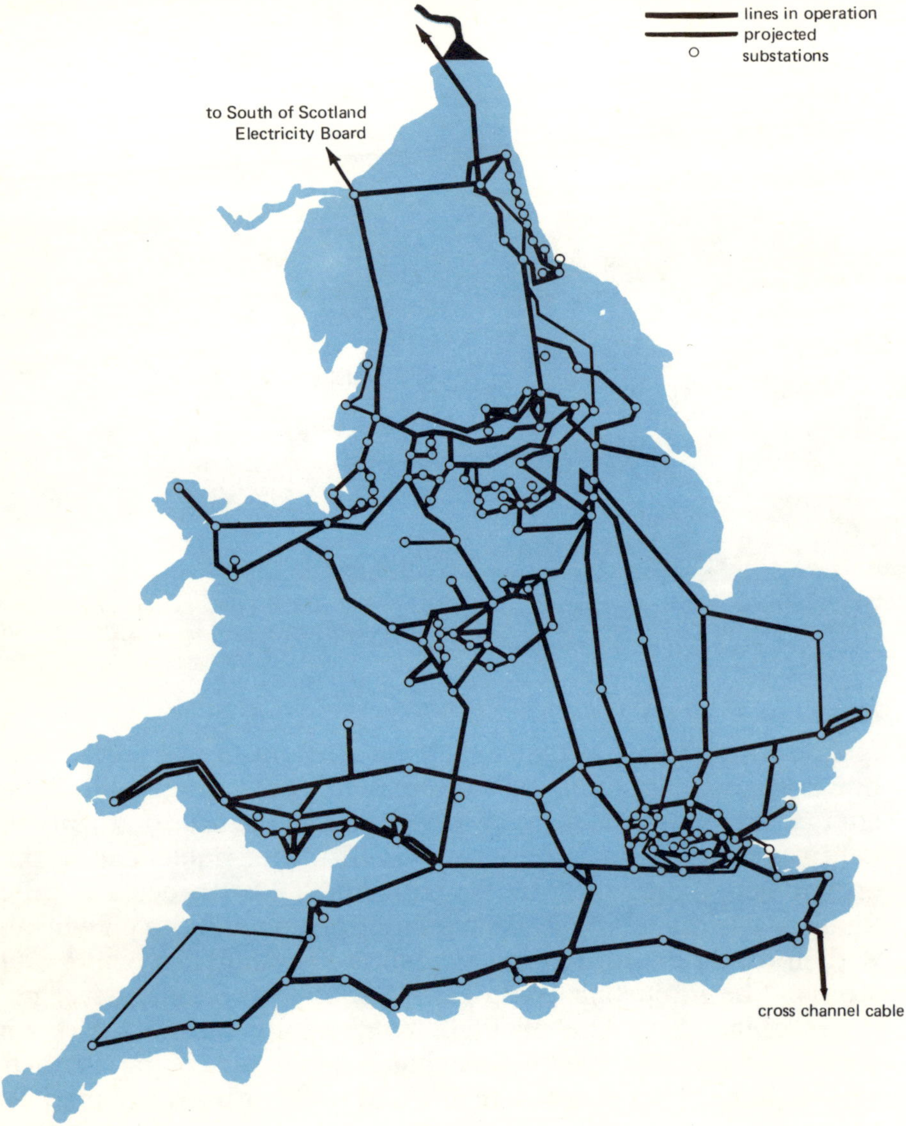

7:2 Simplified diagram showing the main transmission lines of the National Grid.

7:3 Nuclear reactor at Dungeness power station in Kent. The machine on the gantry is recharging the reactor.

Nuclear reactors

In 1945 the British Government decided that a highly industrial nation such as ours could not afford to ignore such an important new source of power as nuclear power, and that every effort should be made to produce cheap electricity in nuclear reactors. Experimental stations were then set up at Harwell in Berkshire and Risley in Derbyshire where experiments were carried on until, in 1953, it was decided to build a reactor at Calder Hall, Cumberland. This was the world's first industrial-sized atomic power station. Since that time improved types of reactors have been built and brought into service with the Central Electricity Generating Board to provide electrical power for us all. Britain now has 29 electricity-producing reactors in operation, and 11 more are being built.

There are four main types of reactors either in service or being used experimentally in Britain now. The earliest are, of course, based on the Calder Hall prototype. The chain reaction from the fuel element is controlled by inserting control rods of the element boron, which absorbs the extra neutrons. The further these rods are lowered into the reactor, the slower the chain reaction becomes. Similarly, the smaller the amount of boron, the faster the reaction and the greater the amount of heat produced. It has also been proved that if the escaping neutrons are slowed down, they can more easily hit and split the nuclei. So to slow down their movement graphite, like the lead in your pencil, is introduced into the reactor. The graphite is called a moderator because it moderates or slows down the neutrons in the chain reaction. The uranium 235, the boron and the graphite together form the nuclear reactor, and because together they produce radioactive material which is highly dangerous to man, they are placed inside a concrete shield.

The high temperatures which the reaction produces are transmitted to a heat exchanger by means of a gas called carbon dioxide. This passes through the reactor to be heated, and then passes into the heat exchanger which contains pipes carrying water. The hot carbon dioxide heats the water as it passes through the exchanger and back to the reactor. The water turns to steam and is fed into the high-pressure turbine, to do just the same job of producing electricity as we read about in Chapter 5 where coal was used to produce the steam.

The next type of reactor, called the advanced gas-cooled reactor (A.G.R.), is very similar to the one just described except for the fuel element. No longer is the fuel natural uranium – it is 'enriched' uranium, made from uranium oxide pellets held in stainless-steel tubes. This improvement means that the fuel has a longer life, and the gas temperatures are allowed to rise much higher, giving greater efficiency all round.

The third type of reactor is called a 'steam generating heavy water reactor' (S.G.H.W.R.), and it is very advanced compared with the two earlier types. The others use graphite as a moderator, but the S.G.H.W.R. uses heavy water, which is made chemically different from the water we drink, and is very good for the job it has to do.

The 'enriched' uranium fuel is contained in tubes surrounded by the heavy water, which in turn is surrounded by ordinary water. As the reactor functions and the heavy water is heated, the ordinary water is circulated round the system until steam is produced. Again the steam is used to power the turbines and generator, to be returned from the condenser and reheated again.

One important by-product of both this reactor and the advanced gas-cooled reactor is the production of plutonium, a radioactive element not found naturally which is heavier and richer as a nuclear fuel than uranium.

Scientists believe that the future of nuclear power rests with the use of plutonium in what are called fast reactors. Britain has a fast reactor at Dounreay in Scotland, which has produced more electricity than any other fast reactor in the world.

The important feature with this reactor, apart from the plutonium fuel, is that it does not use gas or heavy water to transfer the heat from the chain reaction to the ordinary water to make steam. Instead it uses a liquid metal called sodium, which is pumped round the reactor and through two heat exchangers: the first to pre-heat the sodium; the second to heat the ordinary water to create steam.

7:4 Dounreay's experimental fast reactor.

7:5

7:6

7:7

7:8

7:9

7:5 Grid sub-station showing power lines and transformer.

7:6, 7:7 Signs on grid sub-station gates giving information and warning of the great danger to humans.

7:8 Large insulator on high voltage cable.

7:9 Electricity pylon.

7:10 Concrete electric cables sign.

7:10

It is estimated that by the year 2000, Britain will obtain over 75 per cent of her electricity supplies from nuclear power, but at the present time we have many important methods of providing this power. Nuclear power plays a part but so, also, does coal, oil, gas and water. With each power source we have seen that water is heated to make steam, which turns turbines and generators. It is the generator which actually produces the electricity.

The generator

The inventor of this means of producing electricity was Michael Faraday, who in 1831 discovered that electricity could be made to flow through a wire. He moved some wire between the ends of magnets and found that every time he did so a current of electricity flowed along the wire. An efficient generator was not produced, however, until 50 years later when the public began to demand electric lighting and power in their homes, shops and factories.

In the power-station generators, the shaft from the steam turbine is joined to the centre rotating part, which is an electro-magnet made up of many windings of wire. This rotates inside more stationary windings around it, so that electricity is made. In the largest modern generators this electricity is generated at 25,000 volts. A volt is a measure of force in electricity. This voltage is not high enough to transmit the electricity over long distances, so it is passed into a transformer which is able to increase the voltage to 132,000, 275,000 or even 400,000 volts.

7:11 Michael Faraday (1791–1867).

Pylons

The electricity is now ready to be sent into our grid system. This is a network of wires overhead or cables underground which are linked together to cover the whole country with electricity supplies from all our power stations. Because of the very great expense of putting cables under the ground, most of our high-voltage electricity is carried throughout the country along overhead wire supported by huge pylons. The electricity-carrying wires are made of aluminium, strengthened with steel, because electricity passes easily along it. To make the pylons safer because such high-voltage electricity flows between them, they are insulated (protected) by means of strange-shaped glass or porcelain objects called insulators, which you must have observed hanging from the cross-members of the pylons.

Sub-stations – the grid system

Near or in every town or city are located sub-stations which receive the high-voltage electricity and reduce its power, by passing it through transformers, to 33,000 volts for heavy industries, 11,000 volts for light industries and 240 volts for homes and other general uses. Some large industries have their own sub-stations, as do British Rail, who have theirs beside the tracks. Their latest rail-electrification schemes work at 25,000 volts. Please remember that electricity is very dangerous, and even at the house-supply voltage of 240, it can kill or seriously burn you if you are stupid or careless enough to do foolish things with it.

Did you know that we import and export electricity? Between Southern England and France there are two cables under the English Channel so that electricity can be passed between the two countries. At the present time, we use far more French electricity than they do ours. It is certainly very useful to us when the demand for electricity is high in this country to be able to switch into the French grid system and help ourselves.

7:12, 7:13 The pylon having been erected, the riggers are stringing the 400-kV cables and insulators between the pylons.

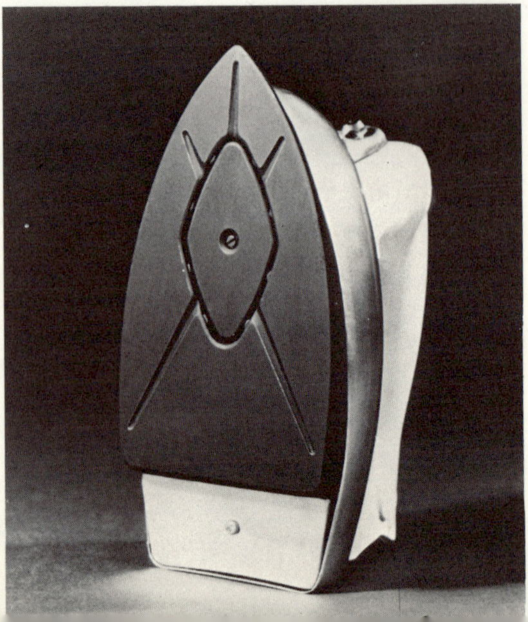

7:14–7:18 What other electrical appliances can you think of that are used around the home?

7:19 With the French coast in the background the cable-laying ship *Ampre* continues its job of connecting the English and French grid systems by means of under-sea cables.

Discovery Assignment 7

1 How does the electricity arrive at your school? The cables could be overhead or underground. If they are overhead cables, how are they different from telephone wires?
2 Where is your nearest sub-station? What information is there supplied on the building or on the gate?
3 How many sub-stations can you locate in your area?
4 Pylons are not pretty sights in our countryside but to our present way of life they are essential. How many different types can you identify among those you see around you or while you are out driving in the country? Make drawings or take photographs of the different types. Electricity is dangerous; never play with it and stay clear of pylons.
5 The riggers, who erect the pylons and span the high-voltage wires between them, have a very dangerous and difficult job to do. Write an account of these dangers and difficulties.
6 If you were a rigger how would you go about spanning a wide river, a motorway or a chasm? Many ingenious methods have been devised. See if you can decide upon some of them yourself.

8-FUTURE POWER SUPPLIES

During the thousands of years man has been on the earth, he has constantly tried to improve his position in the world. By using his superior intelligence he has dominated other creatures so that they produced food and worked for him. When he realised that their strength alone was not enough for his needs, he attempted to conquer nature by harnessing the wind and water to supply the power he needed. He was only partly successful in this venture, but this did not worry him because the power in coal was discovered and put to use. This was followed by even more useful sources of power in the form of gas and oil. Still man remained unsatisfied and searched for more power, until atomic energy was developed. But again we find that our civilisation is continuing to progress, and as it does, so man demands yet more and more power.

The far-sighted scientists of today realise only too well that our future progress depends on the solving of two linked problems. In the past we were fortunate enough not to have to worry about our known sources of power, and the leading scientists of any period in our history have only had to experiment to find new power sources or to make more efficient the power they were using at that time.

Today, however, we have reached, or are very close to reaching, an important point in the development of man. We realise that

8:1 Satellites in space depend very greatly on their solar cells to carry out their function of transmitting information back to earth.

one part of our dual problem is that the supplies of coal, oil and gas which we so greedily consume today, and are continuing to use at an ever-increasing rate, are rapidly running out. Once they have been exhausted, they can never be replaced. The second part of the problem is that, like the men who have gone before us, we are determined to progress, and for this we need new power. These, then, are our problems: we must find new sources of power to replace the coal, gas and oil which we are depleting; and, at the same time, we need a new source of power to ensure man's continuing progress. Never has such a situation faced man before.

Water power

One obvious solution, and this is being attempted, is to improve the other types of power which we already have. Into this group comes water, of which our planet has a plentiful supply. More and more water is being put to use to make electricity in hydro-electric plants. Tremendous theories are being put forward for making vast hydro-electric plants. One such suggestion is to build a barrage, an immense dam-like structure, across the Strait of Gibraltar, the link between the Atlantic Ocean and the Mediterranean Sea. The landlocked Mediterranean loses a lot of water by evaporation, and this water is replaced by water flowing in from the Atlantic Ocean. If a barrage were to be built, it is hoped that this flow of water could be harnessed to provide electricity.

Sea water temperatures

Another water-power theory which is being worked on, although it is not very successful at the moment, is based on the difference in temperature between water from deep in the ocean and water at the surface. A French scientist, Claude Boucherot, designed an ocean power station which sucked warm water from the surface of the ocean into a cylinder from which as much air as possible had been extracted. It has long been known that the lower the pressure of air, the lower will be the boiling point of water. The water in Boucherot's vacuum cylinder boiled, made steam, and turned turbines and generators to produce electricity. The steam was condensed by cold water from the depths of the ocean. Having made a successful working model, he built a full-scale test station in Cuba where the surface water temperature is 28 °C and is only 8 °C at a depth of 600 metres (2,000 ft). His design did produce electricity, but the experiment failed because of the constant trouble he had with the long pipe leading 600 metres into the ocean depths.

More recently another Frenchman, Jules Beau, set up an experimental station off the coast of West Africa, but he was defeated for the same reason as Boucherot. By persisting with this theory it is hoped to produce electricity by this means in the future.

Hydrogen

In the chapter on atomic energy we read how this form of power was first used as a bomb but was later controlled in a reactor to produce electricity. At the present time man has an even more powerful bomb which he is trying to control in a similar way. The hydrogen bomb works by forcing two atoms of hydrogen to join together. When this is done it results in more energy being produced than in any reaction known to man. If the plentiful supply of hydrogen in the sea could be used in a controlled hydrogen reactor, man would have power for millions of years to come.

8:2, 8:3 This is how the Frenchman, Claude Boucherot, set up his experimental station to produce electricity (8:3). From the experimental work done by him and Jules Beau it may be possible to set up power stations in the future like the one in 8:2 to produce electricity.

8:2 FLOATING POWER STATION OF THE FUTURE

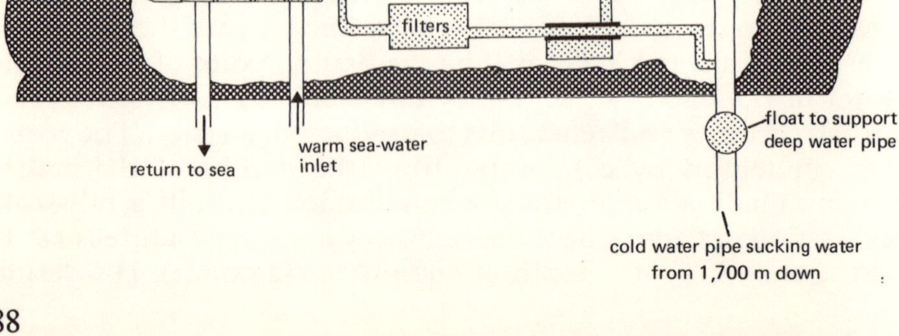

LAND BASED SEA-WATER POWER STATION

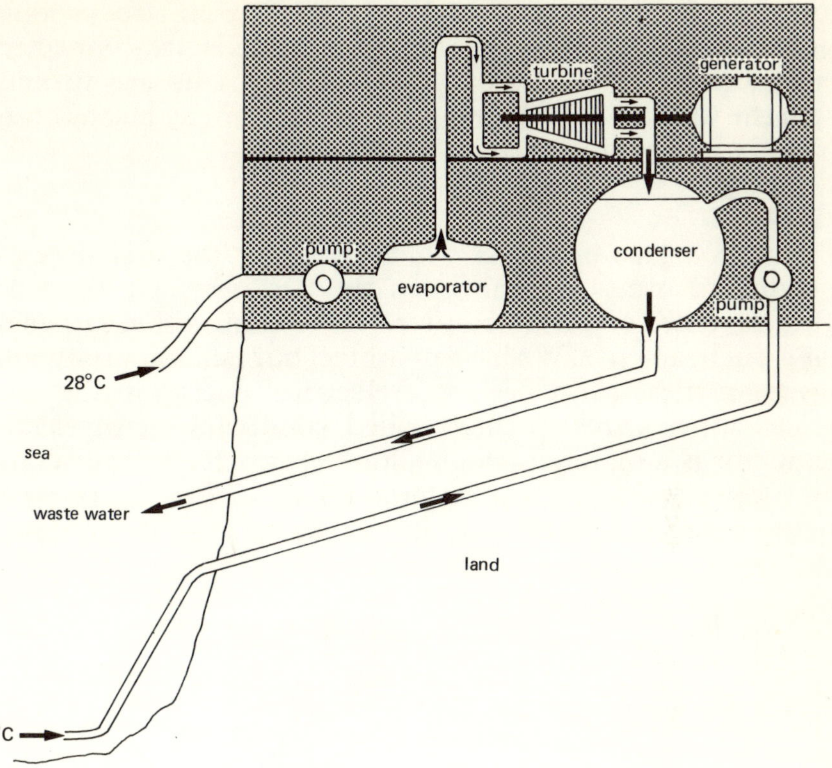

The earth's natural heat

The centre of the earth is thought to be a huge molten ball of iron at a very high temperature, and it has been proved that the deeper the bore-hole into the earth's crust, the higher the temperature becomes. Perhaps some day in the future this supply of power will be tapped and put into the service of man.

Wind

Two other natural sources of power which have been known to us for thousands of years are being experimented with at the moment, but it is difficult to imagine either of them ever really helping with our power problems in a major way. The wind and the energy it contains has been a great servant in the past, but new experiments with highly efficient sails in the state of Vermont, in America, and at Yalta, in Russia, are producing electricity. The sails of these giant windmills are connected to generators to provide this power. Their sails are turned to catch the wind on top of their high steel towers by means of small electric motors.

Lightning

A natural source of power which man, for the first time ever, is trying to use is lightning. In the past we have thought of lightning more of a source of terror and destruction rather than being put to any use. We have to protect our tall, important buildings from these great flashes of electrical energy in the sky by fixing copper wires to them called conductors. (An electrical conductor is a material which allows electricity to pass along it, and copper is a very good material for this.) The lightning conductor attracts the electricity in the sky and lets it run down to

8:4 This dramatic photograph of an electric storm over Johannesburg shows how much energy is being lost. All man can do is to run it to earth through lightning conductors to ensure that his buildings are undamaged.

earth where it can do no harm, rather than strike the building and perhaps cause serious damage.

Weather records prove that in every minute of the day there are about fifty thunderstorms raging somewhere around the world, and all the electrical energy in them is running to earth and to waste. If these powerful currents of electricity could be attracted to special conductors and run down to storage batteries, they could then be passed to transformers which would control the electrical output and transmit it to the grid systems of every nation.

With both modern windmills and lightning 'collectors', the power is of an unreliable source and frequently occurs more where people do not need the power than where they do, so they are not likely to solve our power problems to any great degree.

THE SUN

Perhaps man's greatest hope for power in the future will come from the oldest thing known to us, which also supports our whole way of life, the sun. Some scientists believe that the earth was once part of the sun until ages ago when it was thrown off into space to cool down, and eventually support life which is still nourished by the sun. The sun holds the earth in orbit, gives us light, warmth and our weather and yet it is 149 million kilometres (93 million miles) away. Even at this distance its rays can be used on earth to supply our need for power.

This direct use of the sun's energy has been used for centuries by man in a very limited way. Archimedes is said to have set fire to the sails of the Roman fleet, as they were preparing the invasion, by reflecting the sun's rays. By using magnifying lenses or reflecting mirrors, high temperatures have been achieved in historical times. Lavoisier, the Frenchman, melted platinum in his solar furnace which reached a temperature of nearly 2000 °C, and this was in the 18th century! A solar device at Las Salinas in Chile was built over 90 years ago to distil water at the rate of over 27,000 litres (6,000 gallons) a day.

Solar cooker

The sun's energy can be used to cook food, and this method of cooking is actually in use in countries which have dependable amounts of sunshine, such as parts of India and North Africa.

8:5 Solar reflectors at the solar energy laboratory in France.

Solar furnaces

The big brother of the solar cooker is the solar furnace. This has been developed in America, Algeria and Russia, the latter having one over 9 metres (30 ft) in diameter used to power a steam plant. At the top of the solar furnace league, however, is France. A huge curved mirror has been built on one side of a ten-storey research building. This mirror is curved so that the rays from the sun are reflected to one point near the centre, which means that the heat is concentrated on a small area to achieve the highest possible temperature. In fact this can reach 3,000 °C. Because it is so vast, it can melt huge amounts of metal at a time and is, therefore, of use to French industry as well as being very valuable as a research unit.

Houses can be heated solely by the sun, even in quite cool areas of the world. Large windows collect heat, as well as allowing light to pass through, during the day. This heat is stored in heat-retaining containers so that after the sun has gone down they can be used to warm the house. Offices and schools also can be heated in this way.

8:6 Parabolic mirror which concentrates the sun's rays.

All the uses of the sun's energy so far have been to help warm us and provide energy for industry, but man needs mobility. Nowadays we travel by car, bus, aeroplane, ship and train, or by rocket into space. None of the new sources have given us the power we need for mobility, except that electricity can be used for some cars, trains or trolley buses. The sun can, however, aid mobility through the use of solar batteries. They can be used to power a car, and the American army has a lightweight solar boat in use. At some time in the future this type of solar energy may be developed to maintain our transport system.

Many of our present sources of power are used in such a way that they pollute our world. The advantage of the more recently introduced forms of energy like natural gas and hydro-electricity and of the sources of energy which may be developed in the future, such as hydrogen reactors, seawater power stations and solar energy is that they should cause little or no pollution problems for mankind.

Discovery Assignment 8

1 What will life be like if we do not find sufficient useful forms of power to carry on our way of life when the present sources are depleted?

2 Can you think of new sources of power which could be used and which have not been mentioned in this chapter? For example, do you think that in the next century man will be digging mines and drilling on planets other than this one?

3 Imagine that man was able to control the weather, the wind and the rainfall. Would this be a help in the future for our power supplies or could it cause chaos? Discuss the possibilities with your class.

Books For Further Reading

Adams, H. *Electricity and Its Uses* Blackwell
Baxter, Eric *The Study Book of Coal* The Bodley Head
Coats, Norman *Energy and Power* Weidenfeld & Nicolson
Davey, John *Coal Mining* A. & C. Black
Hammersley, Alan *Weather and Life* Blandford
Hinkelbein, Albert *Energy and Power* Collins
Hogben, Lancelot and Jackson, David *The Wonderful World of Energy* Macdonald
Jones, Eric *Weather and Life* Blandford
Kay, F. G. *It's Made Like This – Electricity* John Baker
Larsen, Egon *Atoms and Atomic Energy* Weidenfeld & Nicolson
Perry, G. A. *Minerals, Mines and Mining* Blandford
Pringle, Patrick *The Conquerors – Power* Max Parrish
Walmsley, Gordon *It's Made Like This – Petrol and Oils* J. Baker
Warburton, C. *The Study Book of Power* The Bodley Head
Wymer, Norman *Behind the Scenes in an Oilfield* J. Baker
Wynyard, John *Dams and Canals* Wheaton
Wynyard, John *Fuel and Power* Macdonald Junior Reference Library

INDEX